Kaushik Kumar, Chikesh Ranjan
Smart Materials

Also of interest

Functional Nanomaterials.
Applications in Medicine and Life Sciences
Maximilian Kryschi, Carola Kryschi and Stefanie Klein, 2024
ISBN 978-3-11-060545-7, e-ISBN 978-3-11-060549-5

Intelligent Materials and Structures.
2nd Edition
Haim Abramovich (Ed.), 2021
ISBN 978-3-11-072669-5, e-ISBN 978-3-11-072670-1

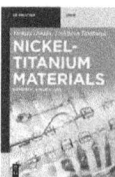

Nickel-Titanium Materials.
Biomedical Applications
Yoshiki Oshida and Toshihiko Tominaga, 2020
ISBN 978-3-11-066603-8, e-ISBN 978-3-11-066611-3

Carbon-Based Smart Materials
Constantinos A. Charitidis, Elias P. Koumoulos
and Dimitrios A. Dragatogiannis (Eds.), 2020
ISBN 978-3-11-047774-0, e-ISBN 978-3-11-047913-3

Kaushik Kumar, Chikesh Ranjan

Smart Materials

Electro-Rheological Fluids, Piezoelectric Smart Materials,
and Shape Memory Alloys

DE GRUYTER

Authors
Dr. Kaushik Kumar
Department of Mechanical Engineering
Birla Institute of Technology
835215 Mesra
Ranchi, Jharkhand
India
kkumar@bitmesra.ac.in

Dr. Chikesh Ranjan
Department of Mechanical Engineering
National Institute of Technology
Rourkela, Odisha 769008, India
j.chikesh123@gmail.com

ISBN 978-3-11-137901-2
e-ISBN (PDF) 978-3-11-137962-3
e-ISBN (EPUB) 978-3-11-137976-0

Library of Congress Control Number: 2024951229

Bibliographic information published by the Deutsche Nationalbibliothek
The Deutsche Nationalbibliothek lists this publication in the Deutsche Nationalbibliografie;
detailed bibliographic data are available on the Internet at http://dnb.dnb.de.

© 2025 Walter de Gruyter GmbH, Berlin/Boston
Cover image: Devrimb/iStock/Getty Images Plus
Typesetting: Integra Software Services Pvt. Ltd.

www.degruyter.com
Questions about General Product Safety Regulation:
productsafety@degruyterbrill.com

Preface

The authors are pleased to present the book *Smart Materials: Electro-Rheological Fluids, Piezoelectric Smart Materials, and Shape Memory Alloys*. The book title was chosen to provide a comprehensive and accessible resource that explores the rapidly evolving field of advanced materials, with a specific focus on smart and innovative materials. This book aims to bridge the gap between scientific research, technological applications, and the broader societal impact of these materials. It aims to empower readers, from students to professionals, with a deep understanding of the principles, properties, and applications of *Smart Materials* toward societal and industrial applications in various industries and domains.

Smart materials are advanced materials that can change their properties in response to external stimuli such as temperature, light, pressure, or electrical fields. These materials exhibit dynamic behaviors, allowing them to adapt to environmental changes, making them useful in a wide range of applications, from aerospace and robotics to healthcare and consumer electronics. Examples of smart materials include shape-memory alloys (SMAs), which can return to a predefined shape when heated, and piezoelectric materials, which generate an electric charge under mechanical stress. Their ability to self-adjust enhances efficiency, functionality, and innovation across industries.

The book delves into the realm of advanced materials engineered with responsive capabilities, adapting their properties in reaction to environmental triggers such as temperature, stress, or electromagnetic fields. This comprehensive exploration covers the design, synthesis, characterization, and applications of these materials, elucidating their role in pioneering technologies such as self-healing structures, sensors, actuators, and energy-efficient systems across diverse sectors like engineering, healthcare, and consumer electronics.

Unveiling the future of materials and exploring the fascinating world of adaptive substances that respond intelligently to external stimuli are the main purposes of the book. From self-repairing structures to cutting-edge electronics, this essential guide illuminates how these materials are revolutionizing industries. Dive in and discover the limitless potential of smart materials.

The book is tailored to cater to a specialized audience comprising researchers, engineers, and professionals deeply entrenched in the fields of materials science, engineering, and related technologies. It delves into intricate concepts and advanced applications, making it an invaluable resource for those seeking an in-depth understanding and exploration of smart materials and their multifaceted applications.

The entire book is segregated into **five chapters** describing the various smart materials that are currently the buzz ones and are quite complicated but very popular among students, researchers, and industrialists and generally included in the course curriculum of undergraduate courses of different universities and institutions. Each

https://doi.org/10.1515/9783111379623-202

chapter ends with a summarization of the content and some review questions so that the reader can assess their understanding.

Chapter 1 provides a comprehensive overview of different classes of materials, including their uses and advancements in intelligent materials. It covers traditional materials such as metals, ceramics, polymers, and composites, emphasizing their unique properties and applications. The chapter also explores intelligent or smart materials that can dynamically respond to external stimuli such as temperature, pressure, and pH, with examples including SMAs and piezoelectric materials. It evaluates the state of materials science, discussing structural materials, functional materials, and polyfunctional materials. Additionally, it delves into the generation and diverse applications of smart materials in fields such as medicine, aerospace, and environmental monitoring, and examines biomimetic materials and their technological impacts.

From this point, the book dives into specific class of smart materials; hence, **Chapter 2**, titled *Smart Materials and Structural Systems*, explores the integration of advanced technologies and materials into intelligent structures. It covers the essential components of smart materials, including thermal materials for heat management and sensing technologies that enable real-time data collection. Microsensors and intelligent systems play a key role in enabling adaptive behaviors, while hybrid smart materials combine multiple properties for enhanced functionality. The chapter outlines an algorithm for synthesizing smart materials, detailing the process of combining base materials and functional components. It distinguishes between passive sensory smart structures, which monitor without active input, and reactive actuator-based systems, which actively respond to stimuli. Additionally, it discusses active sensing and reactive smart structures that provide dynamic adaptation, smart skins for aeroelastic tailoring of aerofoils to optimize aerodynamics, and the synthesis of future smart systems integrating artificial intelligence, sensor networks, and energy-efficient technologies for autonomous and adaptable applications.

The next chapter of the book, **Chapter 3** on *electrorheological (ER) fluids*, explores the fundamental properties and applications of these smart materials that change the viscosity in response to an electric field. It covers the composition of **suspensions and ER fluids,** the **Bingham body model** for describing their flow behavior, and contrasts **Newtonian** and **non-Newtonian viscosity**. Key characteristics of ER fluids, including their field-dependent viscosity, reversibility, and rapid response, are detailed along with the **ER phenomenon** and **charge migration mechanism**. The **ER fluid domain** highlights the various conditions under which ER fluids operate, while **ER fluid actuators** and **design parameters** emphasize their precise control and customization. Applications span across clutches, dampers, brakes, and other systems where ER fluids provide adaptable and efficient performance.

Chapter 4 is on *piezoelectric smart materials*, which explores various phenomena and applications of materials that exhibit piezoelectric, electrostrictive, and pyroelectric properties. Piezoelectric materials, such as PZT (lead zirconate titanate) and PVDF (polyvinylidene fluoride) films, generate an electric charge under mechanical stress, enabling their use in sensors, actuators, and energy-harvesting devices. Electrostriction, which involves the deformation of dielectric materials in response to an electric field, complements piezoelectric effects in precision applications. Pyroelectricity, the generation of charge in response to temperature changes, further extends these materials' utility. Industrial piezoelectric materials are classified based on their properties, such as high sensitivity and stability, with PZT being widely used in actuators and PVDF in flexible sensors. The properties of commercial piezoelectric materials are essential for designing smart devices, from medical diagnostics to wearable tech. Additionally, smart composite laminates with embedded piezoelectric actuators allow for structural health monitoring and adaptive controls. SAW (surface acoustic wave) filters, which leverage piezoelectric effects, are vital in communication systems for filtering specific frequencies. Overall, the chapter covers the principles, properties, and diverse applications of piezoelectric smart materials, emphasizing their integration into advanced technological systems.

The last chapter of the book, **Chapter 5** on *SMAs*, describes the unique materials that can revert to a predetermined shape when subjected to specific stimuli such as temperature changes. These materials, such as nickel alloys and titanium-based nitinol, have characteristics like shape memory and superelasticity due to martensitic and austenitic phase transformations. The thermoelastic martensitic transformation is key to their behavior. Cu-based SMAs are also explored for specialized applications, including aerospace. SMAs find their use in various fields like chirality in advanced materials, fasteners, SMA fibers in reaction vessels, and nuclear reactors. Microrobots and satellite antennas are actuated using SMAs, while blood clot filters showcase the medical potential. Challenges remain, including design impediments and manufacturing limitations such as primary and secondary molding processes. SMA plastics, a growing field, offer innovative applications such as adaptive medical devices, reconfigurable packaging, and smart textiles, broadening the scope of these versatile materials.

In conclusion, this book is a culmination of years of exploration, research, and personal growth. It is our hope that the insights and knowledge shared within these pages will inspire, challenge, and guide you in your own journey. Whether you are a seasoned professional or a curious learner, we believe that the ideas presented here will resonate and offer value. As you turn the pages, we invite you to engage with the material, question assumptions, and explore new perspectives. Thank you for allowing us to share this work with you – we hope it serves as a meaningful contribution to your endeavors.

Kaushik Kumar
Chikesh Ranjan

Acknowledgments

Writing this book has been an incredible journey, and we would not have been able to complete it without the support, guidance, and encouragement of many individuals.

First and foremost, we would like to thank God for providing the opportunity, believe in passion, hard work, and pursue dreams. In the process of putting this book together, it was realized how true this gift of writing is for anyone. This could never have been done without power provided by you and, of course, the faith in You, the Almighty.

We thank our families for having the patience with us for taking yet another challenge that decreases the amount of time we could spend with them. They were our inspiration and motivation, and their constant encouragement and late-night brainstorming sessions were invaluable. The unwavering support and insightful feedback helped to shape the direction of this book, and to say least, their belief in our vision made all the difference.

We would like to thank our parents and grandparents for allowing us to follow our ambitions.

We would like to thank all of our colleagues and friends in different parts of the world for sharing ideas in shaping our thoughts.

We would also like to thank the reviewers, the editorial board members, project development editor, and the complete team of De Gruyter for their constant and consistent guidance, support, and cooperation at all stages of this project. Their effort in every phase, from inception to execution, cannot be expressed in words. Your keen eye and thoughtful suggestions helped in bringing clarity and polish to this manuscript. We deeply appreciate your hard work and dedication, support, and cooperation rendered to the project.

A special thank you to our colleagues and mentors who provided guidance, inspiration, and the occasional push when it was most needed.

Finally, we are grateful to our readers, whose curiosity and passion are the real motivation behind this work. It is our hope that this book inspires, educates, or simply entertains in the way we intended. Our efforts will come to a level of satisfaction if the students, researchers, and professionals concerned with all the fields related to new and smart materials, in particular, and product development, in general, gets benefitted.

Thank you all for being a part of this journey.

Kaushik Kumar
Chikesh Ranjan

https://doi.org/10.1515/9783111379623-203

Contents

1 Introduction and historical perspective

1.1 Introduction

Smart materials are advanced materials engineered to respond dynamically to environmental changes or external stimuli such as temperature, stress, electric or magnetic fields, and pH levels. These materials, such as piezoelectric materials, shape-memory alloys (SMAs), electroactive polymers (EAPs), and thermochromic substances, can significantly alter their properties in a controlled manner. They are utilized in diverse applications like medical devices, robotics, aerospace, and consumer electronics, providing enhanced performance and adaptability. While they drive innovation and offer remarkable benefits, challenges such as high costs, durability issues, and complexity in design and control systems remain [1]. Despite these challenges, the continued development of smart materials promises substantial advancements in technology and various industries.

1.2 Classes of materials and their uses

Materials can be broadly categorized into several classes based on their properties and applications. These classes include metals, ceramics, polymers, composites, and smart materials, as shown in Figure 1.1.

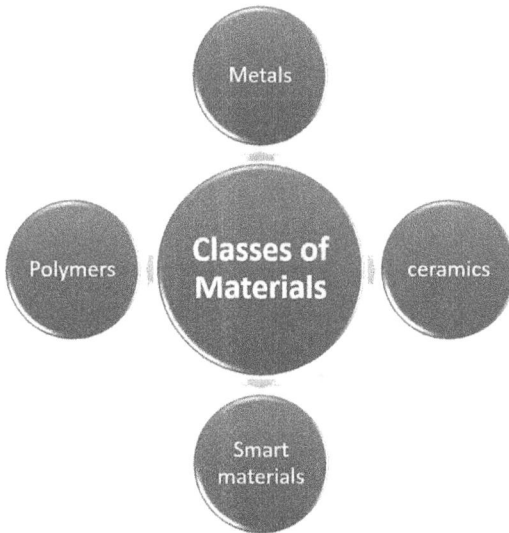

Figure 1.1: Classes of materials.

https://doi.org/10.1515/9783111379623-001

Each class has unique characteristics that make them suitable for specific applications:

– **Metals**:
  – **Properties**: High strength, ductility, conductivity (both thermal and electrical), and malleability.
  – **Usage**: Used in construction (steel beams), transportation (aluminum in aircraft), electronics (copper in wiring), and tools and machinery (stainless steel).
– **Ceramics**:
  – **Properties**: High hardness, brittleness, high melting points, and good thermal and electrical insulation.
  – **Usage**: Employed in applications requiring wear resistance and thermal stability, such as in aerospace (heat shields), medical devices (dental implants), and electronics (insulators).
– **Polymers**:
  – **Properties**: Low density, flexibility, corrosion resistance, and good insulating properties.
  – **Usage**: Found in packaging (plastics), textiles (nylon, polyester), automotive parts (dashboard components), and medical applications (disposable syringes).
– **Composites**:
  – **Properties**: Combination of two or more materials, resulting in enhanced properties such as increased strength, reduced weight, and improved thermal stability.
  – **Usage**: Used in aerospace (carbon fiber reinforced polymers for aircraft structures), sports equipment (carbon fibers in surfboards), and construction (reinforced concrete).
– **Smart materials**:
  – **Properties**: Ability to respond to external stimuli (temperature, stress, and electric or magnetic fields) with a significant change in their properties.
  – **Usage**: Employed in medical devices (SMAs in stents), robotics (EAPs in artificial muscles), aerospace (adaptive structures using piezoelectric materials), and consumer electronics (thermochromic materials in displays).

Each class of material offers distinct advantages that are leveraged across various industries to improve the performance, durability, and functionality of products and systems.

1.3 Intelligent/smart materials

Intelligent or smart materials are advanced materials engineered to respond dynamically to external stimuli or environmental changes. These stimuli can include temperature, stress, electric or magnetic fields, moisture, and pH levels. Smart materials can significantly alter their physical properties in a controlled manner, making them highly adaptable and versatile. Examples include piezoelectric materials that generate electricity under mechanical stress, SMAs that return to a predetermined shape when heated, EAPs that change their shape under an electric field, and thermochromic materials that change their color with temperature [2]. Their applications span across various fields such as medical devices, robotics, aerospace, and consumer electronics, providing innovative solutions for improved performance and functionality.

1.4 Evaluation of materials science

Materials science is a multidisciplinary field that focuses on the properties, performance, and applications of materials. The evaluation of materials science involves a structured process, as illustrated in Figure 1.2, beginning with defining the objective, where the purpose of the material evaluation is clearly identified based on its intended application, such as strength or durability. Following this, material selection is carried out by choosing suitable materials that meet specific requirements, including mechanical, thermal, or electrical properties. Once the materials are selected, the next step is sample preparation, where the material is shaped or cut into appropriate forms for testing. This leads to the experimental testing phase, which involves conducting mechanical, thermal, and electrical tests to assess the material's performance. Data is collected from these tests, such as stress–strain curves or thermal resistance measurements, and then analyzed and interpreted to determine how well the material meets expectations [3]. The material's overall performance is evaluated based on this data, considering factors like strength, stability, and suitability for the intended use. Finally, the results are reported, and decisions are made on whether the material meets the necessary specifications or requires further refinement.

The evaluation of materials science encompasses several key aspects:

– **Historical development**:
  – **Progression**: Materials science has evolved from traditional metallurgy and ceramics to include polymers, composites, and advanced smart materials. The field's growth has been driven by the need for new materials to meet technological advancements and societal needs.
  – **Milestones**: Key historical milestones include the Bronze Age, the Iron Age, the advent of synthetic polymers in the twentieth century, and the development of nanomaterials and smart materials in recent decades.

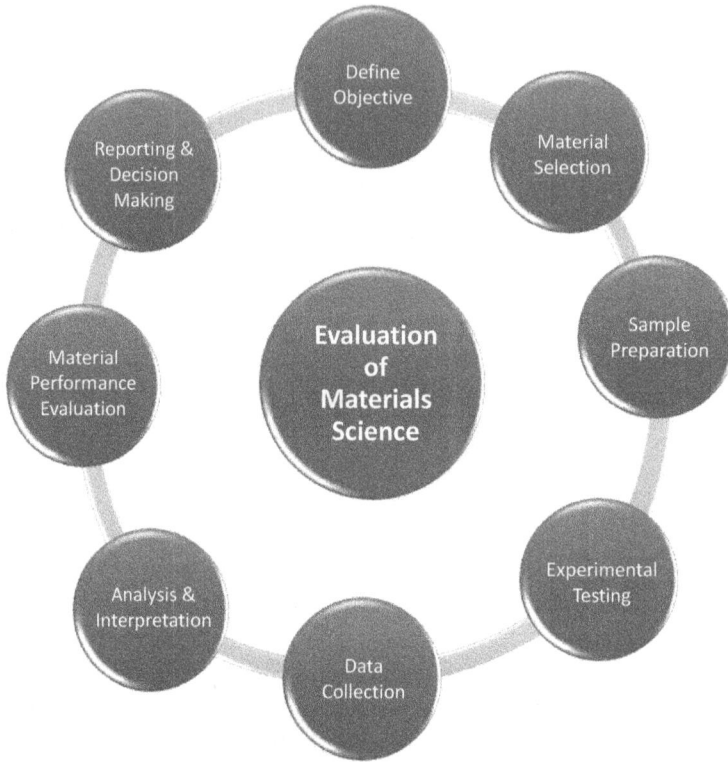

Figure 1.2: Flow diagram of the evaluation of materials science.

- **Key concepts**:
  - **Structure–property relationship**: Understanding how a material's structure at various scales (atomic, microscopic, and macroscopic) influences its properties (mechanical, electrical, and thermal).
  - **Processing–structure relationship**: How different manufacturing and processing techniques affect the material's internal structure and, consequently, its properties and performance.
- **Interdisciplinary nature**:
  - **Integration**: Materials science integrates principles from physics, chemistry, engineering, and biology. This interdisciplinary approach is essential for developing new materials and improving existing ones.
  - **Applications**: The field's applications range from everyday items like plastics and metals to high-tech innovations like semiconductors, biomaterials, and aerospace composites.

- **Technological impact**:
    - **Innovation**: Advances in materials science have been pivotal in the development of new technologies, such as microelectronics, renewable energy systems, and medical devices.
    - **Sustainability**: Materials science also plays a crucial role in addressing environmental challenges by developing sustainable materials and improving recycling processes.
- **Challenges and future directions**:
    - **Sustainability**: Developing materials that are environmentally friendly, recyclable, and sustainable is a significant challenge.
    - **Advanced materials**: The ongoing research aims to create materials with superior properties, such as higher strength-to-weight ratios, better thermal stability, and enhanced electrical conductivity.
    - **Nanotechnology**: The manipulation of materials at the nanoscale offers exciting possibilities for new functionalities and improved performance in various applications.
    - **Cost and scalability:** Making advanced materials affordable and scalable for mass production.
    - **Durability and lifespan:** Ensuring long-term performance without degradation.
    - **Interdisciplinary collaboration:** Combining insights from chemistry, physics, engineering, and biology to innovate and solve complex material challenges.

1.5 Structural materials

Structural materials are materials used in construction and manufacturing that provide the necessary strength, durability, and stability to structures and components [4]. These materials are critical for ensuring the safety and longevity of buildings, bridges, vehicles, and other infrastructure.

1.5.1 Types of structural materials

1.5.1.1 Metals
Metals are a class of materials characterized by their excellent conductivity, strength, and malleability. They are widely used in various industries due to their diverse properties and applications:
- **Properties of metals**
    - **High strength and durability**: Metals are known for their ability to withstand significant forces without breaking or deforming. This makes them ideal for construction, transportation, and machinery.

- **Electrical and thermal conductivity**: Metals are excellent conductors of electricity and heat, making them essential in electrical wiring, electronic components, and heat exchangers.
- **Ductility and malleability**: Metals can be drawn into wires (ductility) or hammered into sheets (malleability) without breaking, allowing for versatile manufacturing processes.
- **Luster**: Metals have a shiny appearance when polished, which is often desirable for decorative and aesthetic purposes.
- **Magnetic properties**: Some metals, such as iron, cobalt, and nickel, exhibit magnetic properties, which are useful in motors and electronic devices.
- **Types of metals**
  A. **Ferrous metals**: They contain iron as the main element.
    - **Example**: Steel.
    - **Properties**: High strength, hardness, and durability.
    - **Usage**: Construction (beams and reinforcements), automotive (frames and engines), and manufacturing (tools and machinery).
  B. **Nonferrous metals**: They do not contain iron, generally more resistant to corrosion.
    - **Example**: Aluminum.
    - **Properties**: Lightweight, corrosion-resistant, and good conductor of electricity.
    - **Usage**: Aerospace (aircraft bodies), transportation (car bodies), and packaging (cans and foils).
  C. **Precious metals**: They are rare metals with high economic value.
    - **Example**: Gold.
    - **Properties**: Excellent corrosion resistance, electrical conductivity, and ductility.
    - **Usage**: Jewelry, electronics (circuit boards and connectors), and financial assets (bullion and coins).
  D. **Alloys**: Mixtures of two or more metals, or metals with other elements, to enhance certain properties.
    - **Example**: Brass (copper and zinc).
    - **Properties**: Improved strength, corrosion resistance, and workability compared to pure metals.
    - **Usage**: Musical instruments, plumbing fittings, and decorative items.
- **Applications of metals**
  - **Construction**: Metals such as steel and aluminum are used extensively in building structures, bridges, and infrastructure due to their strength and durability.
  - **Transportation**: Metals are crucial in manufacturing vehicles, aircraft, ships, and trains. For example, aluminum is used for its lightweight and corrosion resistance in aircraft and automotive industries.

- **Electronics**: Metals like copper and gold are essential for wiring, connectors, and components in electronic devices due to their excellent electrical conductivity.
- **Medical devices**: Metals such as titanium and stainless steel are used in implants, surgical instruments, and medical devices due to their biocompatibility and strength.
- **Consumer goods**: Metals are used in household appliances, cookware, and decorative items for their durability, aesthetic appeal, and functional properties.

1.5.1.2 Concrete

Concrete is a widely used construction material composed of cement, water, aggregates (such as sand, gravel, or crushed stone), and sometimes additives to enhance its properties [5]. It is known for its high compressive strength, durability, and versatility, making it an essential material in construction.

- **Composition of concrete**
    - **Cement**: The binding agent, typically Portland cement, which reacts with water to form a hard, solid matrix.
    - **Water**: Essential for the chemical reaction with cement (hydration) that hardens the mixture.
    - **Aggregates**: Provide volume and contribute to the strength and stability of the concrete.
        - **Fine aggregates**: Sand or crushed stone with smaller particle sizes.
        - **Coarse aggregates**: Gravel or crushed stone with larger particle sizes.
    - **Additives (admixtures)**: Chemicals added to modify specific properties of concrete, such as setting time, workability, and durability.

Examples: Accelerators, retarders, plasticizers, and air-entraining agents.

- **Properties of concrete**
    - **High compressive strength**: Concrete is exceptionally strong under compression, making it ideal for load-bearing applications.
    - **Durability**: Concrete is resistant to weathering, chemical attack, and wear, providing long service life in various environments.
    - **Versatility**: Concrete can be molded into virtually any shape and size, suitable for a wide range of construction projects.
    - **Thermal mass**: Concrete has the ability to absorb and store heat, contributing to energy efficiency in buildings.
    - **Low tensile strength**: Although concrete is strong in compression, it is weak in tension and often needs reinforcement.
- **Types of concrete**
    - **Plain concrete**: Basic mixture of cement, water, and aggregates without reinforcement.

Usage: Sidewalks, pavements, and non-load-bearing structures.
- **Reinforced concrete**: Concrete with embedded steel reinforcement (rebar) to improve tensile strength [11].
  Usage: Structural components such as beams, columns, slabs, and foundations.
- **Precast concrete**: Concrete elements cast and cured in a controlled environment, then transported to the construction site.
  Usage: Prefabricated building components, such as panels, beams, and girders.
- **Prestressed concrete**: Concrete in which internal stresses are introduced to counteract tensile stresses during service.
  Usage: Long-span bridges, high-rise buildings, and heavy-load structures.
- **Lightweight concrete**: Concrete made with lightweight aggregates or foam to reduce density.
  Usage: Insulation, roof decks, and non-load-bearing walls.
- **High-performance concrete**: Designed with superior properties such as high strength, durability, and workability.
  Usage: Specialized structures requiring enhanced performance, such as high-rise buildings and infrastructure projects.
- **Applications of concrete**
  - **Building construction**: Used in foundations, floors, walls, and roofs due to its strength and versatility.
  - **Infrastructure**: Essential for roads, bridges, dams, tunnels, and sewage systems, providing structural integrity and durability.
  - **Residential projects**: Commonly used in driveways, patios, and walkways.
  - **Industrial projects**: Used in factories, warehouses, and large-scale industrial facilities for its ability to withstand heavy loads and harsh conditions.
- **Advantages of concrete**
  - **Strength and durability**: Capable of withstanding heavy loads and harsh environmental conditions.
  - **Versatility**: Can be shaped and molded into various forms to meet different design requirements.
  - **Economical**: Generally cost-effective due to the widespread availability of raw materials.
  - **Fire resistance**: Noncombustible and provides excellent fire resistance, enhancing building safety.
  - **Low maintenance**: Requires minimal upkeep over its lifespan, contributing to lower long-term costs.
- **Disadvantages of concrete**
  - **Low tensile strength**: Requires reinforcement to handle tensile stresses, increasing complexity and cost.

- **Heavy weight**: Can impose significant loads on foundations and supporting structures.
- **Environmental impact**: Cement production contributes to CO_2 emissions, raising concerns about sustainability.
- **Long curing time**: Requires time to achieve full strength, potentially slowing down construction schedules.

1.5.1.3 Wood

Wood is a natural, versatile material derived from trees, widely used in construction, furniture making, and various other applications. It has been a primary building material for centuries due to its abundance, workability, and aesthetic appeal.

- **Properties of wood**
  - **Strength**: Wood exhibits good strength-to-weight ratio, making it capable of supporting significant loads without excessive weight.
  - **Flexibility and workability**: Wood is easy to cut, shape, and join using simple tools, making it ideal for a variety of construction and manufacturing processes.
  - **Thermal insulation**: Wood has excellent insulating properties, providing effective thermal resistance that helps in maintaining energy efficiency in buildings.
  - **Aesthetic appeal**: The natural texture, color, and grain patterns of wood contribute to its visual appeal, making it a preferred material for decorative and architectural elements.
  - **Renewability**: As a renewable resource, wood can be sustainably harvested and regrown, contributing to environmentally friendly building practices.
  - **Hygroscopic nature**: Wood can absorb and release moisture, which affects its dimensional stability and can lead to expansion or contraction.
- **Types of wood**
  - **Softwood**: Derived from coniferous trees (e.g., pine, spruce, and fir).
    Properties: Generally softer and less dense than hardwoods, easy to work with, and typically more affordable.
    Usage: Common in construction (framing and roofing), furniture, and paper products.
  - **Hardwood**: Derived from deciduous trees (e.g., oak, maple, and cherry).
    Properties: Typically harder and denser than softwoods, offering greater strength and durability.
    Usage: High-quality furniture, flooring, cabinetry, and decorative veneers.
  - **Engineered wood**: Manufactured by binding or fixing wood particles or fibers together with adhesives.
    Examples: Plywood, particleboard, medium-density fiberboard, and laminated veneer lumber.

Properties: Enhanced strength, stability, and uniformity compared to natural wood; often more resistant to warping and splitting.

Usage: Widely used in construction (subflooring and wall panels), furniture, and cabinetry.

– **Applications of wood**
  1. **Construction**:
     – **Framing**: Softwood lumber is extensively used in residential and commercial building frameworks.
     – **Roofing and flooring**: Plywood and other engineered wood products are common in roof decking and flooring substrates.
     – **Exterior and interior finishes**: Wood is used for siding, trim, paneling, and flooring, providing both structural and aesthetic benefits.
  2. **Furniture making**:
     – **Custom and mass-produced furniture**: Hardwood and engineered wood are used to make a wide range of furniture, from bespoke pieces to mass-produced items.
     – **Cabinetry**: Both solid wood and engineered wood are used for kitchen and bathroom cabinets due to their durability and workability.
  3. **Decorative elements**:
     – **Millwork**: Includes moldings, trims, and other decorative wood elements that enhance the architectural design of buildings.
     – **Veneers**: Thin slices of wood glued onto core panels (usually of a different material) to create surfaces that look like solid wood.
  4. **Outdoor uses**:
     – **Decking and fencing**: Treated softwood and durable hardwood are used for outdoor structures due to their ability to withstand weather conditions.
     – **Landscaping**: Wood is used in pergolas, gazebos, garden furniture, and other outdoor features.

– **Advantages of wood**
  – **Sustainability**: Wood is a renewable resource, and sustainable forestry practices can ensure a continuous supply.
  – **Energy efficiency**: Wood structures often require less energy to build compared to other materials, and their insulating properties can reduce energy costs.
  – **Biodegradability**: Wood is biodegradable, reducing its environmental impact at the end of its life cycle.
  – **Natural beauty**: Wood's natural appearance can enhance the aesthetic value of structures and objects.

– **Disadvantages of wood**
  – **Susceptibility to pests and decay**: Wood can be vulnerable to insects, fungi, and rot, requiring treatments and maintenance.

- **Dimensional instability**: Wood can expand or contract with changes in moisture, leading to warping or cracking.
- **Flammability**: Wood is a combustible material, which necessitates proper fire safety measures in construction.
- **Variability**: The natural variability in wood can lead to inconsistencies in appearance and performance.

1.5.1.4 Polymers

Polymers are large molecules composed of repeated subunits called monomers, which are chemically bonded together. These materials are known for their versatility, as their properties can be tailored to meet a wide range of applications. Polymers can be natural, such as cellulose and proteins, or synthetic, such as plastics and nylons.

- **Properties of polymers**
    - **Lightweight**: Polymers are generally lighter than metals and ceramics, making them ideal for applications where weight reduction is crucial.
    - **Corrosion resistance**: Most polymers are resistant to chemical attacks and do not corrode, unlike metals.
    - **Flexibility**: Polymers can be flexible and elastic, making them suitable for products that need to bend or stretch.
    - **Insulating properties**: Polymers often have excellent thermal and electrical insulating properties.
    - **Variable mechanical properties**: The strength, toughness, and hardness of polymers can be adjusted through the choice of monomers and polymerization processes.
- **Types of polymers**
    - **Thermoplastics**: These polymers can be melted and reshaped multiple times without significantly altering their chemical properties.
    Examples: Polyethylene (PE), polypropylene, polystyrene (PS), and polyvinyl chloride (PVC).
    Usage: Packaging (plastic bags and bottles), automotive parts, toys, and household items.
    - **Thermosetting polymers**: Once cured or hardened, these polymers cannot be remelted. They form a rigid, inflexible structure.
    Examples: Epoxy, phenolic, and melamine.
    Usage: Adhesives, coatings, electrical insulators, and composite materials.
    - **Elastomers**: Polymers with elastic properties, allowing them to return to their original shape after being stretched or compressed.
    Examples: Natural rubber and synthetic rubber (e.g., neoprene and silicone).
    Usage: Tires, seals, gaskets, and flexible hoses.
    - **Biodegradable polymers**: Designed to decompose naturally by the action of living organisms, typically bacteria.

Examples: Polylactic acid (PLA) and polyhydroxyalkanoates.

Usage: Medical implants, packaging, agricultural films, and disposable items.

- **Applications of polymers**
    - **Packaging**: Polymers are extensively used in packaging due to their lightweight, flexibility, and ability to form airtight seals.

    Examples: Plastic bags, bottles, films, and containers.
    - **Automotive and aerospace**: Polymers help reduce weight and improve fuel efficiency while providing durability and resistance to corrosion.

    Examples: Dashboard components, bumpers, fuel tanks, and interior trims.
    - **Construction**: Polymers are used for insulation, piping, coatings, and structural components.

    Examples: PVC pipes, insulation foams, and roofing membranes.
    - **Electronics**: Polymers are used as insulators, in flexible circuits, and as protective casings.

    Examples: Circuit boards, connectors, and housing for electronic devices.
    - **Medical devices**: Biocompatible and biodegradable polymers are used in a variety of medical applications.

    Examples: Surgical sutures, implants, drug delivery systems, and prosthetics.
    - **Textiles**: Synthetic fibers made from polymers are used in clothing, upholstery, and industrial fabrics.

    Examples: Nylon, polyester, and spandex.
- **Advantages of polymers**
    - **Versatility**: Polymers can be engineered to exhibit a wide range of properties, making them suitable for diverse applications.
    - **Cost-effective**: Many polymers are inexpensive to produce and process.
    - **Durability**: Polymers are resistant to corrosion, chemicals, and environmental degradation.
    - **Lightweight**: Their low density makes them ideal for applications where weight savings are important.
    - **Ease of manufacture**: Polymers can be molded, extruded, and cast into complex shapes with relative ease.
- **Disadvantages of polymers**
    - **Environmental impact**: Many polymers, especially traditional plastics, are not biodegradable and contribute to pollution and landfill waste.
    - **Thermal sensitivity**: Some polymers can degrade at high temperatures, limiting their use in high-heat applications.
    - **Mechanical limitations**: While polymers can be strong and tough, they often do not match the strength and stiffness of metals or ceramics.
    - **UV degradation**: Some polymers can degrade when exposed to ultraviolet light over long periods.

1.5.1.5 Ceramics

Ceramics are a diverse class of materials known for their hardness, brittleness, heat resistance, and electrical insulation properties. They are typically made from inorganic, nonmetallic materials that are processed and then fired at high temperatures to achieve their final form. Ceramics have been used for thousands of years, and modern advancements have expanded their applications in various industries.

- Properties of ceramics
  - **High hardness and strength**: Ceramics are extremely hard and can withstand significant mechanical stresses, making them ideal for applications where wear resistance is crucial.
  - **Brittleness**: While ceramics are strong under compression, they are brittle and can fracture under tension or impact.
  - **High melting points**: Ceramics can withstand very high temperatures without melting or degrading, making them suitable for high-temperature applications.
  - **Electrical insulation**: Most ceramics are excellent electrical insulators, which is why they are often used in electronic and electrical applications.
  - **Chemical inertness**: Ceramics are resistant to many chemicals and do not corrode easily, which makes them ideal for use in harsh chemical environments.
  - **Low thermal conductivity**: Many ceramics are good thermal insulators, which are beneficial in applications requiring heat resistance.
- Types of ceramics
  - **Traditional ceramics**:
    Examples: Clay, porcelain, and earthenware.
    Usage: Commonly used in pottery, bricks, tiles, and sanitary ware.
    Properties: Made from natural materials like clay and feldspar; typically processed by shaping and firing.
  - **Advanced ceramics**:
    Examples: Alumina (Al_2O_3), silicon carbide (SiC), and zirconia (ZrO_2).
    Usage: Used in high-tech applications such as electronics, aerospace, medical devices, and cutting tools.
    Properties: Manufactured with high purity and controlled composition; exhibit superior mechanical, thermal, and electrical properties.
  - **Glass ceramics**:
    Examples: Pyroceram and Foturan.
    Usage: Used in cookware, optical components, and dental restorations.
    Properties: A hybrid of glass and ceramic properties; formed by controlled crystallization of glass.

- **Applications of ceramics**
  - **Construction**:
    Bricks and tiles: Used in building construction for walls, roofs, and flooring due to their durability and aesthetic appeal.
    Sanitary ware: Toilets, sinks, and bathtubs made from ceramic materials are resistant to water and stains.
  - **Electronics and electrical engineering**:
    Insulators: Used in power transmission lines and electronic circuits for their excellent electrical insulation properties.
    Capacitors and piezoelectric devices: Used in electronic components that require stable and reliable performance.
  - **Aerospace and automotive**:
    Thermal barriers: Used in jet engines and exhaust systems to protect components from high temperatures.
    Wear-resistant parts: Engine components, such as bearings and seals, are made from ceramics for their durability and resistance to wear.
  - **Medical applications**:
    Implants: Used in hip replacements and dental implants for their biocompatibility and wear resistance.
    Bone grafts and prosthetics: Ceramics like hydroxyapatite are used to repair and replace the bone due to their similarity to natural bone minerals.
  - **Industrial and cutting tools**:
    Cutting tools: Advanced ceramics are used in machining and cutting operations due to their hardness and ability to retain sharp edges.
    Industrial components: Wear-resistant components such as nozzles, bearings, and seals in various industrial applications.
- **Advantages of ceramics**
  - **High-temperature stability**: Suitable for applications involving extreme temperatures.
  - **Hardness and wear resistance**: Ideal for cutting tools, protective coatings, and applications requiring durability.
  - **Corrosion resistance**: Performs well in chemically aggressive environments.
  - **Electrical insulation**: Provides excellent insulation in electrical and electronic applications.
  - **Biocompatibility**: Suitable for medical implants and devices.
- **Disadvantages of ceramics**
  - **Brittleness**: Prone to fracture and chipping under stress or impact.
  - **Difficult to machine**: Hardness makes ceramics difficult to cut, shape, and finish, often requiring specialized equipment.
  - **Complex manufacturing process**: High-temperature processing and sintering can be complex and costly.
  - **Limited tensile strength**: Not suitable for applications involving tensile loads.

Structural materials play a crucial role in ensuring the integrity and functionality of various structures and components, contributing to safety, performance, and longevity in construction and manufacturing.

1.6 Functional materials

Functional materials are engineered to possess specific properties and perform particular functions beyond their basic structural roles. These materials are designed to interact with their environment or respond to external stimuli such as temperature, pressure, electric or magnetic fields, light, and chemical environments. They play a crucial role in advanced technologies and are integral to various innovative applications in multiple industries.

- **Properties of functional materials**
  - **Responsive behavior**: They have the ability to change their properties or behaviors in response to external stimuli.
  - **Tailored properties**: They are designed to exhibit specific physical, chemical, electrical, or optical properties for specialized applications.
  - **Integration capability**: They can be integrated into devices and systems to enhance performance and functionality.
  - **Multifunctionality**: They are often capable of performing multiple functions simultaneously, such as sensing and actuating.
- **Types of functional materials**
  - **Smart materials**:

Smart materials, also known as intelligent or responsive materials, are materials that can change their properties or behavior in response to external stimuli such as temperature, pressure, electric or magnetic fields, light, moisture, or chemical environments. These materials are designed to mimic the adaptive capabilities found in natural systems and can be integrated into a variety of applications to enhance functionality and performance.

- **Properties of smart materials**
  - **Responsiveness:** They have the ability to react to specific external stimuli predictably.
  - **Reversibility:** Many smart materials can return to their original state once the stimulus is removed.
  - **Sensitivity:** They have high sensitivity to even small changes in the external environment.
  - **Adaptability:** They can adapt their properties to changing conditions, improving performance and functionality.

- **Types of smart materials**
  - **SMAs:** Metals that return to a predefined shape when heated above a certain temperature.
    Example: Nickel–titanium (nitinol).
    Usage: Actuators, medical devices (stents and guidewires), and eyeglass frames.
  - **Piezoelectric materials:** Generate an electric charge in response to mechanical stress or, conversely, change the shape when an electric field is applied.
    Examples: Lead zirconate titanate (PZT) and quartz.
    Usage: Sensors, actuators, microphones, and vibration control systems.
  - **Thermochromic materials:** Change color in response to temperature changes.
    Examples: Liquid crystals and vanadium dioxide.
    Usage: Thermometers, smart windows, mood rings, and temperature indicators.
  - **Photochromic materials:** Change color when exposed to light.
    Examples: Silver chloride and spiropyrans.
    Usage: Sunglasses, UV-sensitive coatings, and optical data storage.
  - **Electrochromic materials:** Change color or opacity when an electric current is applied.
    Examples: Tungsten oxide and polyaniline (PANI).
    Usage: Smart windows, displays, and rearview mirrors in cars.
  - **Magnetorheological (MR) and electrorheological (ER) fluids:** Change viscosity in response to magnetic or electric fields, respectively.
    Examples: Suspensions of magnetic particles in a carrier fluid (MR fluids); dielectric particles in a nonconducting fluid (ER fluids).
    Usage: Dampers, clutches, and prosthetic devices.
  - **Self-healing materials:** Automatically repair damage without human intervention.
    Examples: Polymers with microencapsulated healing agents, and concrete with bacteria that precipitate calcium carbonate.
    Usage: Coatings, structural materials, and composites.
  - **Shape-memory polymers (SMPs):** Polymers that return to their original shape when triggered by an external stimulus, such as heat.
    Examples: Polyurethane and PLA.
    Usage: Biomedical devices, aerospace components, and textiles.
- **Applications of smart materials**
  - **Medical devices:**
    Smart implants: SMAs are used in stents and orthodontic devices.
    Wearable sensors: Piezoelectric materials are used for monitoring vital signs.
  - **Aerospace and automotive:**
    Adaptive structures: Piezoelectric materials and SMAs for morphing wings and active vibration control [5].

    Smart windows: Electrochromic materials for controlling light transmission and improving energy efficiency.
- **Consumer electronics:**
    Flexible displays: Electrochromic and thermochromic materials for e-readers and smartwatches.
    Responsive surfaces: Touch-sensitive surfaces and haptic feedback devices using piezoelectric materials.
- **Construction:**
    Self-healing concrete: Enhances the durability and lifespan of structures.
    Smart glass: Thermochromic and electrochromic windows for energy-efficient buildings.
- **Textiles:**
    Smart fabrics: SMPs and thermochromic materials for clothing that adapt to environmental conditions.
    Wearable technology: Integrated sensors and actuators for health monitoring and interactive garments.

- **Advantages of smart materials**
  - **Enhanced functionality:** Provide additional capabilities such as self-healing, shape change, and responsiveness to environmental changes.
  - **Energy efficiency:** Can improve energy efficiency in applications such as smart windows and adaptive building materials.
  - **Customization:** Tailored properties for specific applications, leading to improved performance and user experience.
  - **Innovation potential:** Enable the development of new technologies and applications across various industries.
- **Disadvantages of smart materials**
  - **Cost:** Often more expensive to produce and implement compared to traditional materials.
  - **Complex manufacturing**: Requires advanced manufacturing techniques and expertise.
  - **Durability:** Some smart materials may have limited lifespan or stability under certain conditions.
  - **Integration challenges:** Difficulty in integrating with existing systems and materials.
    - **Conductive polymers**:

Conductive polymers are a class of polymers that conduct electricity, combining the electrical properties of metals with the processing advantages and mechanical properties of plastics. Unlike traditional insulating polymers, conductive polymers can support and transport charge through their molecular structure. These materials have opened up new possibilities in electronics, sensors, and various other fields due to their unique properties.

- **Properties of conductive polymers**
  - **Electrical conductivity:** Conductive polymers can conduct electricity due to the delocalization of π-electrons along their backbone. Their conductivity can range from semiconducting to highly conductive, comparable to metals.
  - **Flexibility:** They retain the flexible, lightweight, and processable characteristics of conventional polymers.
  - **Chemical stability:** Many conductive polymers are chemically stable and resistant to environmental degradation, though stability can vary with the specific polymer and environmental conditions.
  - **Tunability:** The electrical properties of conductive polymers can be tailored through chemical synthesis and doping (adding dopants to increase conductivity).
  - **Biocompatibility:** Some conductive polymers are biocompatible, making them suitable for biomedical applications.
- **Types of conductive polymers**
  - **PANI:**
    Properties: Good environmental stability, can be doped to enhance conductivity.
    Usage: Antistatic coatings, corrosion protection, sensors, and flexible electronics.
  - **Polypyrrole (PPy):**
    Properties: High conductivity, ease of synthesis, and biocompatibility.
    Usage: Biosensors, medical devices, and actuators.
  - **Polythiophene:**
    Properties: High conductivity, flexibility, and stability.
    Usage: Organic photovoltaics (OPVs), field-effect transistors, and light-emitting diodes (LEDs).
  - **Polyacetylene:**
    Properties: One of the first discovered conductive polymers, highly conductive when doped.
    Usage: Mainly of historical interest; laid the groundwork for the development of other conductive polymers.
  - **Poly(3,4-ethylene dioxythiophene) (PEDOT):**
    Properties: Excellent conductivity and transparency, especially when combined with polystyrene sulfonate (PEDOT:PSS).
    Usage: Transparent electrodes, touch screens, antistatic coatings, and organic electronics.
- **Applications of conductive polymers**
  - **Electronics:**
    Flexible electronics: Used in flexible circuits, displays, and wearable devices due to their flexibility and conductivity.

Organic light-emitting diodes (OLEDs): Serve as conductive layers in OLED displays and lighting.

OPVs: Used in solar cells to convert light into electricity.

– **Sensors:**

Chemical sensors: Detect gases and vapors through changes in electrical resistance.

Biosensors: Used in medical diagnostics to detect biomolecules, leveraging their biocompatibility and conductive properties.

– **Energy storage:**

Batteries: Used as electrode materials in rechargeable batteries.

Supercapacitors: Serve as active materials in supercapacitors, providing high capacitance and rapid charge/discharge cycles [2].

– **Antistatic and electromagnetic interference (EMI) shielding:**

Antistatic coatings: Applied to surfaces to prevent the buildup of static electricity.

EMI shielding: Used in electronic enclosures to block electromagnetic interference.

– **Biomedical applications:**

Neural interfaces: Used in devices that interface with the nervous system, such as neural electrodes.

Drug delivery: Conductive polymers can be used in systems for controlled drug release.

– **Corrosion protection:**

Coatings: Applied to metal surfaces to prevent corrosion by providing a protective conductive layer.

– **Advantages of conductive polymers**
  – **Flexibility and processability:** Can be processed into various shapes and forms, including films, fibers, and coatings.
  – **Lightweight**: Offer a lightweight alternative to metals and other conductive materials.
  – **Tunability**: Electrical and mechanical properties can be tailored through chemical modifications and doping.
  – **Biocompatibility**: Suitable for use in biomedical applications.
– **Disadvantages of conductive polymers**
  – **Lower conductivity compared to metals**: While conductive, they generally do not match the high conductivity of metals.
  – **Stability issues**: Some conductive polymers may degrade over time, particularly under environmental stress or harsh conditions.
  – **Processing challenges**: This can require specialized processing techniques to achieve desired properties and performance.

- **Cost**: Can be more expensive than traditional polymers, particularly for high-performance variants.
  - **Biomaterials**:
  Biomaterials are substances engineered to interact with biological systems for medical purposes, either therapeutic (treating, augmenting, or replacing tissues) or diagnostic [1]. These materials can be derived from natural sources or synthesized in the lab and must be biocompatible, meaning they should not elicit an adverse reaction when introduced into the body.
- **Properties of biomaterials**
  - **Biocompatibility**: They have the ability to perform with an appropriate host response in a specific situation.
  - **Biodegradability**: For temporary applications, materials can be designed to degrade within the body over time.
  - **Mechanical properties**: Adequate strength, flexibility, and durability to match the biological tissues they replace or support.
  - **Nontoxicity**: They should not release harmful substances into the body.
  - **Bioactivity**: They can interact with biological tissues to promote specific responses, such as cell attachment and proliferation.
- **Types of biomaterials**
  - **Polymers:**
  Natural polymers: Collagen, gelatin, alginate, and chitosan.
  Synthetic polymers: PLA, polyglycolic acid (PGA), and polyethylene glycol.
  Usage: Drug delivery systems, tissue engineering scaffolds, sutures, and wound dressings.
  - **Metals:**
  Examples: Titanium, stainless steel, and cobalt–chromium alloys.
  Usage: Orthopedic implants (hip and knee replacements), dental implants, and cardiovascular stents.
  Properties: High strength, durability, and resistance to wear and corrosion.
  - **Ceramics:**
  Examples: Hydroxyapatite, bioglass, zirconia, and alumina.
  Usage: Bone grafts, dental implants, and coatings for metal implants.
  Properties: High compressive strength, bioactivity, and similarity to natural bone minerals.
  - **Composites:**
  Examples: Mixtures of polymers, ceramics, and metals.
  Usage: Bone and dental implants, tissue scaffolds, and orthopedic devices.
  Properties: Combine the advantages of different materials, such as strength, bioactivity, and flexibility.
  - **Natural biomaterials:**
  Examples: Silk, collagen, cellulose, and chitin.

Usage: Tissue engineering scaffolds, wound healing products, and drug delivery systems.

Properties: Good biocompatibility and biodegradability, often promote cell attachment and growth.

- **Applications of biomaterials**
  - **Medical implants:**

    Orthopedic implants: Joint replacements (hip and knee), bone plates, and screws [1].

    Dental implants: Tooth replacements and dental prosthetics.

    Cardiovascular devices: Stents, heart valves, and vascular grafts.
  - **Tissue engineering:**

    Scaffolds: 3D structures that support cell growth and tissue formation.

    Skin grafts: For burn victims and wound healing.

    Organ regeneration: Efforts to grow functional tissues and organs in the lab.
  - **Drug delivery systems:**

    Controlled release: Polymers that release drugs at a controlled rate.

    Targeted delivery: Systems designed to deliver drugs to specific cells or tissues.

    Biodegradable carriers: Polymers that degrade after delivering their therapeutic load.
  - **Wound healing:**

    Dressings: Materials that promote healing and protect the wound.

    Hydrogels: Provide a moist environment to aid in healing.

    Antimicrobial materials: Prevent infection at the wound site.
  - **Diagnostic devices:**

    Biosensors: Devices that detect biological molecules or changes.

    Imaging agents: Materials that are used in medical imaging to enhance contrast.
- **Advantages of biomaterials**
  - **Enhancing patient outcomes:** Improved functionality and compatibility with the body lead to better recovery and performance of medical devices.
  - **Versatility**: Can be designed and tailored for a wide range of medical applications.
  - **Innovation**: Enable the development of new therapies and treatments.
  - **Longevity**: Many biomaterials are designed to last long term within the body without degradation or loss of function [1].
- **Disadvantages of biomaterials**
  - **Biocompatibility issues**: Potential for immune reactions or rejection by the body.

- **Complex manufacturing**: Requires precise and often costly production processes.
- **Degradation**: Some materials may degrade too quickly or release harmful byproducts.
- **Regulatory hurdles**: Strict regulations for medical devices and implants can delay development and increase costs.

1.7 Polyfunctional materials

Polyfunctional materials, also known as multifunctional materials, are engineered to perform multiple functions simultaneously or sequentially within a single material system. These materials are designed to address complex requirements by integrating various properties and functionalities, thereby enhancing performance, reducing weight, and minimizing the need for multiple distinct materials.

- **Properties of polyfunctional materials**
  - **Multifunctionality:** They are capable of performing multiple roles such as structural support, energy storage, sensing, actuation, and self-healing.
  - **Integration capability:** They combine various functional properties in a single material, facilitating simpler and more efficient system designs.
  - **Adaptive behavior:** They respond to changes in the environment or operating conditions by adjusting their properties.
  - **Enhanced performance:** They have improved overall performance by leveraging the synergies between different functionalities.
- **Types of polyfunctional materials**
  - **Structural composites**: Materials that combine mechanical strength with other functionalities such as thermal management or energy storage.
    Examples: Carbon fiber composites with embedded sensors for structural health monitoring.
    Usage: Aerospace, automotive, and civil engineering applications.
  - **Smart materials:** Materials that can respond to external stimuli and change their properties accordingly.
    Example: SMAs that can also conduct electricity.
    Usage: Actuators, sensors, and adaptive structures.
  - **Self-healing materials**: Materials that can automatically repair damage, restoring functionality.
  - **Energy-harvesting materials**: Materials that can convert environmental energy into electrical energy.
    Example: Piezoelectric materials that also serve as structural components.
    Usage: Wearable electronics, remote sensors, and low-power devices.
  - **Conductive polymers**: Polymers that can conduct electricity while maintaining flexibility and other polymeric properties.

Example: PANI or PEDOT:PSS integrated into flexible electronics that also serve as sensors or actuators.

Usage: Flexible displays, smart textiles, and biomedical devices.

- **Applications of polyfunctional materials**
  - **Aerospace and automotive:**
    Lightweight structures: Composite materials that provide structural integrity, thermal management, and health monitoring.
    Smart components: Materials that can change the shape or properties in response to environmental conditions.
  - **Medical devices:**
    Implants: Materials that combine biocompatibility, structural support, and drug delivery capabilities.
    Wearable health monitors: Flexible materials that can monitor vital signs, provide feedback, and adapt to the body's movements.
  - **Energy systems:**
    Battery components: Electrodes that combine high energy density with mechanical strength and flexibility.
    Solar panels: Materials that integrate light absorption, charge transport, and mechanical support.
  - **Construction and infrastructure:**
    Self-healing concrete: Concrete that can repair its own cracks, enhancing longevity and reducing maintenance costs.
    Smart windows: Windows that adjust their transparency in response to light and temperature, while also providing insulation.
  - **Consumer electronics:**
    Flexible devices: Materials that combine electronic functions with mechanical flexibility and durability.
    Wearable technology: Smart textiles that integrate sensing, actuation, and data transmission capabilities.
- **Advantages of polyfunctional materials**
  - **Reduced weight and volume:** Integrating multiple functions into a single material reduces the need for multiple components, leading to lighter and more compact designs.
  - **Enhanced performance:** Synergistic effects between different functions can lead to improved overall performance.
  - **Simplified design and manufacturing:** Reduces complexity by combining multiple functionalities into fewer materials, streamlining production processes.
  - **Cost efficiency**: Potentially lowers costs by reducing the number of materials and components needed.

- **Disadvantages of polyfunctional materials**
  - **Complexity in design:** Designing materials that effectively combine multiple functions can be challenging.
  - **Manufacturing challenges:** Advanced processing techniques may be required to achieve the desired multifunctional properties.
  - **Potential trade-offs:** Balancing different functionalities may require compromises, potentially limiting the performance of individual properties.
  - **Reliability and durability:** Ensuring long-term reliability and durability of all integrated functions can be difficult.

1.8 Generation of smart materials

Smart materials, also known as intelligent or responsive materials, are materials that can respond to external stimuli and change their properties in a controlled manner. Figure 1.3 outlines the progression of smart materials through four generations, starting with passive smart materials, which respond to external stimuli without active control (e.g., SMAs). The second generation, known as active smart materials, can sense and actively respond to environmental changes (e.g., piezoelectric materials). The third generation, known as multifunctional and adaptive smart materials, can perform multiple functions and adapt to different stimuli, such as self-healing polymers. The most advanced, fourth generation, known as intelligent and autonomous materials, integrate sensing, actuation, and decision-making, enabling autonomous responses without external control. After these materials are developed, their performance is evaluated and tested, followed by real-world applications and optimization for enhanced functionality.

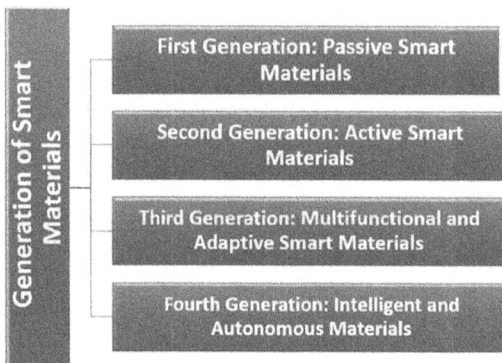

Figure 1.3: Flow diagram of the generation of smart materials.

The development of smart materials has evolved over several generations, each marked by advancements in material science, technology, and application.
– **First generation: passive smart materials**

Characteristics:
– Respond to external stimuli without feedback control.
– Limited to one-time or simple responses.

Examples:
1. **SMAs:**
    – **Function:** Return to their original shape when heated above a certain temperature.
    – **Usage**: Medical devices (stents and orthodontic wires), actuators, and eyeglass frames.
    – **Material**: Nickel–titanium (nitinol).
2. **Piezoelectric materials:**
    – **Function**: Generate an electric charge in response to mechanical stress or change the shape when an electric field is applied.
    – **Usage**: Sensors, actuators, and vibration control systems.
    – **Material**: PZT and quartz.
3. **Magnetostrictive materials:**
    – **Function**: Change the shape or dimensions in response to a magnetic field.
    – **Usage:** Actuators and sensors in sonar systems.
    – **Material**: Terfenol-D.

– **Second generation: active smart materials**

Characteristics:
– Incorporate feedback control mechanisms.
– Capable of continuous or repeated responses to stimuli.

Examples:
1. **EAPs:**
    – **Function:** Change the shape or size in response to an electric field.
    – **Usage**: Artificial muscles, haptic feedback devices, and flexible electronics.
    – **Materials:** PANI and polyvinylidene fluoride (PVDF).
2. **Electrochromic materials:**
    – **Function:** Change the color or opacity when an electric current is applied.
    – **Usage:** Smart windows, displays, and rearview mirrors in cars.
    – **Materials:** Tungsten oxide and PANI.

3. **Photomechanical materials:**
    -   **Function:** Change the shape or mechanical properties in response to light.
    -   **Usage:** Light-driven actuators and switches.
    -   **Materials:** Azobenzene-containing polymers.

Third generation: multifunctional and adaptive smart materials

Characteristics:
-   Combine multiple functions into a single material system.
-   Capable of complex, adaptive, and autonomous responses.

Examples:
1. **Self-healing materials:**
    -   **Function:** Automatically repair damage to restore functionality.
    -   **Usage:** Coatings, structural materials, and electronic devices.
    -   **Material:** Polymers with embedded microcapsules containing healing agents.
2. **Multifunctional composites:**
    -   **Function:** Provide structural support along with additional functionalities such as sensing, actuation, or energy storage.
    -   **Usage:** Aerospace, automotive, and civil engineering applications.
    -   **Materials:** Carbon fiber composites with embedded sensors.
3. **Energy-harvesting materials:**
    -   **Function:** Convert environmental energy into electrical energy.
    -   **Usage:** Wearable electronics, remote sensors, and low-power devices.
    -   **Materials:** Piezoelectric or thermoelectric materials.
4. **Biomimetic materials:**
    -   **Function:** Mimic natural processes or structures to achieve superior functionality.
    -   **Usage:** Medical implants, adaptive camouflage, and soft robotics.
    -   **Materials:** Hydrogels that mimic the properties of natural tissues.

Fourth generation: intelligent and autonomous materials

Characteristics:
-   Capable of self-learning and adapting to changing conditions.
-   Integrated with advanced computing and sensing capabilities.

Examples:
1. **Self-sensing and self-actuating materials:**
   - **Function**: Integrate sensing and actuation capabilities for real-time response and adaptation.
   - **Usage**: Structural health monitoring, adaptive structures, and robotics.
   - **Materials**: Composites with embedded piezoelectric sensors and actuators.
2. **Programmable materials:**
   - **Function:** Change properties or behavior based on preprogrammed instructions or algorithms.
   - **Usage:** 4D printing, where objects change the shape over time or in response to stimuli.
   - **Materials:** SMPs and hydrogels.
3. **Neuromorphic materials:**
   - **Function:** Mimic the neural processes of the brain to achieve cognitive functions.
   - **Usage:** Artificial intelligence, adaptive control systems, and advanced robotics.
   - **Materials:** Organic electronics and memristive materials.

The generation of smart materials has progressed from simple, passive materials to sophisticated, multifunctional, and adaptive systems. Each generation has brought advancements in material science and technology, enabling new applications and improving the functionality and efficiency of existing ones. The future of smart materials lies in their ability to integrate with advanced computing and sensing technologies, leading to intelligent and autonomous systems that can learn, adapt, and evolve over time.

1.9 Diverse areas of intelligent materials

Intelligent materials, also known as smart materials, are designed to respond to external stimuli in a controlled and predictable manner. These materials find applications across various fields due to their unique properties and capabilities. Below are some of the diverse areas where intelligent materials are making significant impacts.

1. Aerospace and defense

Applications:
- **Adaptive structures:** SMAs and piezoelectric materials are used to create structures that can change the shape or stiffness in response to external conditions, improving aerodynamics and structural performance [5].
- **Vibration control:** Piezoelectric and magnetostrictive materials are used in vibration damping systems to protect sensitive equipment and enhance stability.

– **Health monitoring:** Embedded sensors in composite materials monitor the integrity of aircraft structures in real time, allowing for predictive maintenance and enhanced safety.

Examples:
– Morphing wings that adjust their shape for optimal performance.
– Vibration dampers in helicopters using piezoelectric actuators.

2. Biomedical and healthcare

Applications:
– **Drug delivery systems:** Polymers that respond to pH or temperature changes release drugs at controlled rates, enhancing treatment effectiveness.
– **Tissue engineering:** Scaffolds made from biocompatible smart materials support cell growth and tissue regeneration [12].
– **Wearable devices:** Smart textiles embedded with sensors monitor vital signs and other health parameters in real time.

Examples:
– Hydrogels that deliver insulin in response to glucose levels.
– Self-healing materials for wound dressings that promote faster recovery.

3. Automotive industry

Applications:
– **Crash protection:** Materials that absorb the impact energy improve passenger safety.
– **Adaptive headlights:** Materials that change their optical properties based on lighting conditions to provide better visibility.
– **Structural health monitoring:** Embedded sensors in vehicle components monitor stress and strain, alerting to potential failures.

Examples:
– SMAs in adaptive suspension systems.
– Electrochromic windows that adjust tint to reduce glare.

4. Consumer electronics

Applications:
- **Flexible displays**: Organic LEDs (OLEDs) and conductive polymers enable the creation of flexible, foldable screens.
- **Haptic feedback**: Materials that respond to electrical stimuli provide tactile feedback in touchscreens and wearable devices.
- **Energy harvesting**: Piezoelectric materials convert mechanical energy from movements into electrical energy to power small devices.

Examples:
- Smartwatches with flexible displays.
- Smartphones with haptic feedback for enhanced user experience.

5. Construction and infrastructure

Applications:
- **Self-healing concrete**: Materials that can repair cracks autonomously extend the lifespan of structures and reduce maintenance costs.
- **Smart windows**: Electrochromic materials adjust transparency based on sunlight intensity, improving energy efficiency.
- **Structural health monitoring**: Sensors embedded in building materials monitor stress, temperature, and other parameters to ensure structural integrity.

Examples:
- Bridges with embedded sensors for real-time health monitoring.
- Buildings with smart windows that reduce cooling and heating costs.

6. Environmental and energy

Applications:
- **Solar energy harvesting**: Photovoltaic materials convert sunlight into electricity with high efficiency.
- **Energy storage**: Smart materials enhance the performance of batteries and supercapacitors by improving energy density and charge/discharge rates.
- **Water purification**: Responsive materials that change properties to capture and release contaminants.

Examples:
- Perovskite solar cells with high conversion efficiency.
- Batteries with solid electrolytes for improved safety and performance.

7. Robotics and automation

Applications:
- **Soft robotics**: Elastomers and other flexible materials enable robots to perform delicate tasks and interact safely with humans.
- **Actuators and sensors**: Materials that change the shape or properties in response to stimuli drive movement and provide feedback for robotic systems.
- **Artificial muscles**: EAPs that mimic the behavior of biological muscles provide more natural movement for robots.

Examples:
- Robotic grippers with soft, adaptive materials for handling fragile objects.
- Wearable exoskeletons with artificial muscles to assist mobility.

8. Textiles and fashion

Applications:
- **Smart textiles**: Fabrics embedded with sensors and actuators monitor the body temperature, movement, and other parameters for fitness and health applications.
- **Adaptive clothing**: Materials that change properties, such as color or insulation, in response to environmental conditions.
- **Self-cleaning fabrics**: Materials that repel water and dirt, reducing the need for washing.

Examples:
- Athletic wear that monitors the heart rate and hydration levels.
- Jackets that adjust insulation based on the external temperature.

Intelligent materials are revolutionizing numerous industries by providing advanced functionalities and improving performance, efficiency, and user experience. Their ability to respond dynamically to environmental changes opens up new possibilities for innovation and application across diverse fields. As research and development in this area continue to advance, the potential for smart materials to transform existing technologies and create new ones will only grow.

1.10 Primitive functions of intelligent materials

Intelligent materials, or smart materials, possess the unique ability to respond to external stimuli in a controlled and predictable manner. The fundamental or primitive functions of these materials form the basis for their advanced applications across various industries. Here are the key primitive functions of intelligent materials:

1. Sensing

Function:
The ability to detect changes in the environment or external stimuli such as temperature, pressure, light, magnetic fields, and chemical compositions.
Examples:
– **Piezoelectric materials**: Generate an electric charge in response to mechanical stress.
– **Thermochromic materials:** Change color in response to temperature variations.
– **pH-sensitive polymers:** Change their properties in response to changes in the pH level of their environment.

2. Actuation

Function:
The ability to change shape, size, or other physical properties in response to an external stimulus.
Examples:
– **SMAs:** Return to a predefined shape when heated above a certain temperature.
– **EAPs:** Change the shape or size when an electric field is applied.
– **Magnetostrictive materials:** Change their shape in response to a magnetic field.

3. Energy conversion

Function:
The ability to convert one form of energy into another, often used for energy harvesting or generating power from environmental sources.
Examples:
– **Photovoltaic materials:** Convert light energy into electrical energy.
– **Piezoelectric materials:** Convert mechanical energy into electrical energy, and vice versa.
– **Thermoelectric materials:** Convert temperature differences into electrical voltage.

4. Self-healing

Function:
The ability to repair damage autonomously, restoring functionality and extending the material's lifespan.
Examples:
- **Polymers with embedded microcapsules**: Release healing agents when cracks form, bonding the material back together.
- **SMPs:** Recover their original shape after damage when exposed to heat or other stimuli.
- **Metal alloys with healing agents:** Utilize phase transformations to repair micro-cracks.

5. Adaptive response

Function:
The ability to change properties in response to environmental conditions, optimizing performance dynamically.
Examples:
- **Electrochromic materials**: Change opacity in response to electric current, used in smart windows.
- **Photoresponsive polymers:** Alter their physical properties or shape in response to light.
- **Hydrogels:** Expand or contract in response to changes in moisture, temperature, or pH.

6. Controlled release

Function:
The ability to release substances in a controlled manner in response to specific stimuli, commonly used in drug delivery systems.
Examples:
- **pH-responsive polymers**: Release drugs in response to the pH levels in different parts of the body.
- **Temperature-sensitive polymers:** Release therapeutic agents at certain temperatures.
- **Magnetically responsive materials:** Release encapsulated substances when exposed to a magnetic field.

7. Memory

Function:
The ability to remember a predefined shape or configuration and return to it upon stimulation.
Examples:
–   **SMAs**: Recall and return to a specific shape after deformation when heated.
–   **SMPs:** Recover their original shape after being stretched or deformed when exposed to certain temperatures.

8. Damping

Function:
The ability to dissipate energy, usually mechanical vibrations or shocks, to reduce amplitude and prevent damage.
Examples:
–   **MR fluids**: Change viscosity in response to a magnetic field, used in adaptive damping systems.
–   **Viscoelastic materials**: Absorb and dissipate energy from vibrations or impacts, used in shock absorbers.

The primitive functions of intelligent materials provide the foundation for their diverse applications across various fields, from aerospace and automotive to biomedical and consumer electronics. These functions – sensing, actuation, energy conversion, self-healing, adaptive response, controlled release, memory, and damping – enable smart materials to interact dynamically with their environment, leading to innovations that enhance performance, safety, and efficiency. As research continues, these primitive functions will be further refined and expanded, paving the way for even more advanced and versatile applications.

1.11 Intelligent inherent in materials

The concept of inherent intelligence in materials refers to the intrinsic ability of certain materials to sense, respond, and adapt to environmental stimuli without the need for external control systems. These materials possess built-in mechanisms that enable them to perform specific functions autonomously, leveraging their molecular structure and physical properties. Below are key characteristics and examples illustrating how intelligence can be inherent in materials:

1. Self-sensing

Characteristics:
Materials that can detect changes in their environment or internal state without the need of external sensors.

Examples:
- **Piezoelectric materials**: They generate an electrical signal in response to mechanical stress. This property is inherent in the crystal structure of materials like quartz and PZT, making them useful for pressure sensors and accelerometers.
- **Thermochromic materials**: They change color with temperature changes due to intrinsic molecular changes. These are used in applications like temperature indicators and mood rings.

2. Self-actuation

Characteristics:
Materials that can change their shape, size, or other properties in response to external stimuli such as heat, electric fields, or magnetic fields.

Examples:
- **SMAs**: Materials such as nitinol can "remember" their original shape and return to it when heated. The mechanism is due to phase transformations between martensite and austenite phases.
- **EAPs**: They change their dimensions when subjected to an electric field due to the movement of ions or the reorientation of dipoles within the polymer matrix.

3. Self-healing

Characteristics:
Materials that can autonomously repair damage without external intervention.

Examples:
- **Polymers with microcapsules**: They contain healing agents that are released when the material is damaged. The healing process is initiated by the breakage of the microcapsules and subsequent polymerization of the healing agents.
- **Metals and alloys**: Some can undergo self-healing at high temperatures due to the diffusion of atoms that close microcracks.

4. Adaptive response

Characteristics:
Materials that can modify their properties in response to environmental changes, such as humidity, temperature, light, or pH.
Examples:
–   **Hydrogels**: They can swell or contract in response to changes in pH or temperature. This is due to the hydrophilic or hydrophobic interactions within the polymer network.
–   **Photoresponsive polymers:** They change their shape or mechanical properties in response to light exposure. The mechanism involves photoisomerization, where the absorption of light leads to a change in molecular structure.

5. Energy conversion

Characteristics:
Materials that can convert one form of energy into another inherently due to their physical or chemical structure.
Examples:
–   **Photovoltaic materials**: For example, silicon in solar cells can convert light energy directly into electrical energy through the photovoltaic effect.
–   **Thermoelectric materials**: They convert temperature differences into electrical voltage due to the Seebeck effect, where different materials generate a voltage when subjected to a thermal gradient.

6. Controlled release

Characteristics:
Materials that can release substances in a controlled manner in response to specific triggers.
Examples:
–   **pH-responsive polymers:** They release drugs in response to the pH level in different parts of the body. The polymer matrix swells or degrades, releasing the encapsulated drug.
–   **Temperature-sensitive polymers:** For example, poly(N-isopropylacrylamide) (PNIPAM) releases drugs when the temperature reaches a certain threshold due to the polymer's phase transition.

7. Memory and programmability

Characteristics:
Materials that can "remember" previous states or configurations and return to them upon stimulation.
Examples:
– **SMPs**: They return to a preset shape when heated above their glass transition temperature. The shape memory effect is due to the reversible phase transition between the amorphous and crystalline phases of the polymer.
– **MR fluids:** They change their viscosity in response to a magnetic field, which can be used to create programmable damping systems.

Intelligent inherent materials leverage their intrinsic properties to perform complex functions autonomously. These materials are fundamentally designed or naturally endowed with capabilities such as self-sensing, self-actuation, self-healing, adaptive response, energy conversion, controlled release, and memory. The development and application of such materials span numerous fields, including aerospace, biomedical, automotive, consumer electronics, and more, driving innovation and creating smarter, more efficient systems and devices. As research progresses, the potential for new materials with even greater inherent intelligence continues to expand, opening up new possibilities for technology and industry.

1.12 Examples of materials

1.12.1 Intelligent materials

Intelligent materials, also known as smart materials, are engineered to possess intrinsic properties that enable them to respond to external stimuli in a controlled, reversible, and predictable manner. These materials can sense changes in their environment, process the information, and react accordingly. This built-in "intelligence" allows them to perform functions such as self-healing, sensing, actuation, and energy conversion. Intelligent materials are revolutionizing various fields by providing advanced solutions that enhance performance, efficiency, and functionality.

Key characteristics of intelligent materials
– **Responsiveness**:
    – They have the ability to react to external stimuli such as temperature, pressure, light, pH, magnetic fields, and electrical fields.

- **Adaptability:**
  - They have the capacity to adjust properties or behavior in response to environmental changes.
- **Reversibility:**
  - They have the ability to return to the original state once the external stimulus is removed.
- **Integration:**
  - They combine multiple functions (e.g., sensing and actuation) within a single material system.

Types and examples of intelligent materials

1. **SMAs**
   - **Function:**
     - Return to a predefined shape when exposed to a specific temperature.
   - **Example:**
     - Nickel–titanium (nitinol) is used in medical stents, actuators, and eyeglass frames.
2. **Piezoelectric materials**
   - **Function:**
     - Generate an electric charge in response to mechanical stress or change shape when an electric field is applied.
   - **Example:**
     - PZT is used in sensors, actuators, and energy-harvesting devices.
3. **EAPs**
   - **Function:**
     - Change shape or size when subjected to an electric field.
   - **Example:**
     - PANI and PVDF are used in artificial muscles, flexible electronics, and haptic devices.
4. **Magnetostrictive materials**
   - **Function:**
     - Change shape or dimensions in response to a magnetic field.
   - **Example:**
     - Terfenol-D is used in sonar systems, precision actuators, and vibration control.
5. **Thermochromic materials**
   - **Function:**
     - Change color in response to temperature changes.
   - **Example:**
     - Liquid crystal mixtures are used in thermometers, mood rings, and thermal mapping.

6. **Electrochromic materials**
   - **Function:**
     - Change color or transparency when an electric current is applied.
   - **Example:**
     - Tungsten oxide is used in smart windows, displays, and rearview mirrors.
7. **Self-healing materials**
   - **Function:**
     - Repair damage autonomously to restore functionality.
   - **Example:**
     - Polymers with embedded microcapsules containing healing agents are used in coatings, structural materials, and electronic devices.
8. **Photovoltaic materials**
   - **Function:**
     - Convert light energy into electrical energy.
   - **Example:**
     - Silicon is used in solar panels and organic photovoltaic materials in flexible solar cells.
9. **Thermoelectric materials**
   - **Function:**
     - Convert temperature differences into electrical voltage.
   - **Example:**
     - Bismuth telluride is used in thermoelectric generators for waste heat recovery.
10. **Hydrogels**
    - **Function:**
      - Expand or contract in response to changes in moisture, temperature, or pH.
    - **Example:**
      - PNIPAM is used in drug delivery systems, tissue engineering, and sensors.

Applications of intelligent materials
1. **Aerospace and defense**
   - Adaptive structures using SMAs and piezoelectric materials.
   - Vibration damping systems with magnetostrictive materials.
   - Structural health monitoring with embedded sensors.
2. **Biomedical and healthcare**
   - Drug delivery systems using pH-responsive polymers.
   - Tissue engineering scaffolds with biocompatible hydrogels.
   - Wearable health monitoring devices with smart textiles.
3. **Automotive industry**
   - Adaptive suspension systems with SMAs.

- Smart windows with electrochromic materials.
- Energy harvesting from vibrations using piezoelectric materials.

4. **Consumer electronics**
   - Flexible displays with EAPs and conductive polymers.
   - Haptic feedback in touchscreens using piezoelectric materials.
   - Energy harvesting in wearable devices using thermoelectric materials.

5. **Construction and infrastructure**
   - Self-healing concrete to reduce maintenance costs.
   - Smart windows for energy-efficient buildings.
   - Structural health monitoring with embedded sensors.

6. **Environmental and energy**
   - High-efficiency solar cells with photovoltaic materials.
   - Thermoelectric generators for waste heat recovery.
   - Water purification systems with responsive membranes.

Advantages of intelligent materials

- **Enhanced functionality**: Combining multiple functions into a single material system.
- **Improved performance:** Adaptive responses lead to better performance under varying conditions.
- **Reduced weight and complexity:** Integration of multiple capabilities can reduce the need for separate components.
- **Increased safety and reliability:** Real-time monitoring and self-healing capabilities enhance safety and longevity.

1.13 Structural materials

Structural materials are materials that are used to support and transmit loads in buildings, bridges, vehicles, and other structures. They are selected based on their mechanical properties, durability, and cost. The primary function of structural materials is to withstand forces such as tension, compression, shear, and bending without failing. Here are the main types of structural materials along with their characteristics, applications, and examples:

Types and examples of structural materials

1. Metals

Metals are widely used in construction and manufacturing due to their high strength, ductility, and ability to withstand significant loads. They are typically classified into ferrous and nonferrous metals.

- **Ferrous metals:** They contain iron. Examples include steel and cast iron:
    - **Steel**: High tensile strength, durability, and ductility. It is used in construction beams, bridges, and reinforcing bars in concrete.
    - **Cast iron**: Good compressive strength and wear resistance. It is used in heavy machinery bases and some architectural applications.
- **Nonferrous metals**: They do not contain iron. Examples include aluminum and titanium:
    - **Aluminum**: Lightweight, corrosion-resistant, and has good thermal conductivity. It is used in aircraft structures, window frames, and lightweight vehicle components.
    - **Titanium**: High strength-to-weight ratio and excellent corrosion resistance. It is used in aerospace applications and high-performance sporting equipment.

2. Concrete

Concrete is a composite material made from cement, aggregates (such as sand and gravel), and water. It is widely used in construction for its compressive strength, versatility, and durability.
- **Characteristics**:
    - High compressive strength.
    - Low tensile strength (often reinforced with steel bars or fibers).
    - Can be molded into various shapes and sizes.
    - Durable and fire-resistant.
- **Applications**:
    - Building foundations, columns, and beams.
    - Bridges, dams, and tunnels.
    - Pavements, sidewalks, and driveways.

3. Wood

Wood is a natural, renewable material that has been used in construction for thousands of years. It is valued for its aesthetic qualities, ease of use, and good strength-to-weight ratio.
- **Characteristics**:
    - Good tensile and compressive strength.
    - Lightweight and easy to work with using basic tools.
    - Renewable and environmentally friendly.
    - Can be treated to resist decay and insects.
- **Applications**:
    - Residential construction (framing, flooring, and roofing).
    - Furniture and cabinetry.
    - Structural components in bridges and commercial buildings.
    - Decorative elements and finishes.

4. Polymers

Polymers, including plastics and synthetic fibers, are increasingly used in structural applications due to their versatility, lightweight, and resistance to corrosion and chemicals.

- **Characteristics:**
  - Lightweight and strong.
  - Resistant to corrosion, chemicals, and moisture.
  - Can be easily molded and fabricated.
  - Some polymers can be reinforced with fibers (composites) to enhance strength.
- **Applications:**
  - Building insulation, piping, and window frames.
  - Automotive components (bumpers and dashboards).
  - Aerospace components (composite structures).
  - Marine applications (boat hulls and decking).

5. Ceramics

Ceramics are inorganic, nonmetallic materials that are typically hard, brittle, and resistant to heat and chemicals. They are used in structural applications where high temperature and wear resistance are required.

- **Characteristics:**
  - High compressive strength.
  - Excellent thermal and chemical resistance.
  - Brittle and can crack under tensile stress.
  - Durable and wear-resistant.
- Applications:
  - Structural tiles and bricks in buildings.
  - Refractory linings in furnaces and kilns.
  - Wear-resistant surfaces and components in machinery.
  - Insulating components in electrical systems.

1.13.1 Electrical materials

Electrical materials are specialized materials used to conduct, insulate, and store electrical energy in a variety of applications. These materials are critical for the functioning of electrical and electronic devices, ranging from simple household wiring to complex integrated circuits in computers. The main categories of electrical materials include conductors, insulators, semiconductors, and magnetic materials.

Types and examples of electrical materials

1. Conductors

Conductors are materials that allow the easy flow of electrical current due to their low electrical resistance. They are primarily used in wiring and components that require efficient electrical conduction.

- **Characteristics:**
  - High electrical conductivity.
  - Low electrical resistance.
  - Generally metals.
- **Examples:**
  - **Copper**: The most commonly used conductor in electrical wiring due to its excellent conductivity, ductility, and ease of use.
  - **Aluminum**: Used in power transmission lines and overhead cables due to its lower density and cost compared to copper.
  - **Silver**: Has the highest electrical conductivity of all metals but is expensive, so it is used in high-performance applications like connectors and circuit boards.

2. Insulators

Insulators are materials that resist the flow of electric current, providing protection and insulation in electrical systems. They are used to isolate conductive parts and prevent accidental contact with live wires.

- **Characteristics:**
  - High electrical resistance.
  - Low electrical conductivity.
  - Often materials with high dielectric strength.
- **Examples:**
  - **Plastic (PVC and PTFE (polytetrafluoroethylene))**: Commonly used in insulating electrical wires and cables.
  - **Glass**: Used in high-voltage insulators and electronic components.
  - **Ceramics**: Employed in insulators for high-voltage applications and electronic substrates.
  - **Rubber**: Used in insulating gloves, mats, and other protective gear.

3. Semiconductors

Semiconductors have electrical conductivity between that of conductors and insulators. Their conductivity can be manipulated by doping with other elements, making them essential in electronic devices and integrated circuits.

- **Characteristics:**
  - Intermediate electrical conductivity.
  - Ability to control conductivity through doping.
  - Fundamental to modern electronics.
- **Examples:**
  - **Silicon**: The most widely used semiconductor material in electronics, forming the basis of most integrated circuits and transistors.
  - **Germanium**: Used in early transistors and some specialized applications.
  - **Gallium arsenide (GaAs)**: Used in high-speed electronics and optoelectronic devices such as LEDs and laser diodes.

4. Magnetic materials

Magnetic materials are used in applications that require the manipulation of magnetic fields, such as in transformers, inductors, magnetic storage, and electric motors.
- **Characteristics:**
  - Ability to generate or respond to magnetic fields.
  - Divided into ferromagnetic, paramagnetic, and diamagnetic categories.
- **Examples:**
  - **Iron and silicon steel**: Used in transformer cores and electric motor laminations due to their high magnetic permeability and low energy loss.
  - **Ferrites**: Ceramic compounds that are used in high-frequency inductors and transformers.
  - **Alnico**: A type of alloy used in permanent magnets for electric motors and sensors.

5. Dielectric materials

Dielectric materials are insulators that can be polarized by an electric field, making them useful in capacitors and other applications where energy storage and insulation are required.
- **Characteristics:**
  - High dielectric constant.
  - Ability to store electrical energy.
  - High electrical resistance.
- **Examples:**
  - **Ceramic dielectrics**: Used in capacitors and electronic substrates.
  - **Polymer dielectrics (PE and PTFE)**: Used in high-frequency capacitors and as insulation in cables.
  - **Glass**: Used in capacitors and insulating substrates.

6. Piezoelectric materials
Piezoelectric materials generate an electric charge in response to mechanical stress and are used in sensors, actuators, and transducers.
- **Characteristics:**
  - Ability to convert mechanical energy to electrical energy and vice versa.
  - Used in precision applications.
- **Examples:**
  - **Quartz**: Used in oscillators and frequency control devices.
  - **PZT**: Used in ultrasonic transducers and piezoelectric actuators.
  - **PVDF**: Used in flexible sensors and actuators.

Applications of electrical materials
- **Power generation and distribution:**
  - Conductors (copper and aluminum) for wiring and transmission lines.
  - Insulators (ceramics and glass) for high-voltage insulators.
  - Magnetic materials (iron and ferrites) for transformers and inductors.
- **Electronics and computing:**
  - Semiconductors (silicon and GaAs) for integrated circuits, transistors, and diodes.
  - Dielectrics (ceramics and polymers) for capacitors and insulating substrates.
  - Piezoelectric materials (quartz and PZT) for sensors and oscillators.
- **Telecommunications:**
  - Conductive materials (copper and silver) for wiring and connectors.
  - Dielectrics (polymers) for cable insulation and capacitors.
  - Semiconductors (silicon and GaAs) for signal processing components.
- **Automotive and aerospace:**
  - Conductors (aluminum and copper) for wiring harnesses and power systems.
  - Insulators (plastic and ceramics) for protecting electrical systems.
  - Magnetic materials (iron and alnico) for motors and sensors.
- **Consumer electronics:**
  - Semiconductors (silicon) for processors and memory chips.
  - Conductive polymers and flexible materials for touchscreens and wearable devices.
  - Dielectrics (ceramics) for compact capacitors.

1.13.2 Biocompatible materials

Biocompatible materials are designed to interact with biological systems without eliciting any adverse effects, making them crucial in medical and healthcare applications. These materials are used to create medical devices, implants, prosthetics, drug delivery systems, and tissue engineering scaffolds. The primary requirement for a material to be biocompatible is that it must not induce a significant immune response, toxicity, or other negative reactions when in contact with body tissues and fluids.

Characteristics of biocompatible materials
- **Nontoxic:** Must not release harmful substances that can cause damage to tissues or organs.
- **Non-immunogenic:** Should not trigger an immune response that could lead to inflammation or rejection.
- **Durable and stable:** Must maintain integrity and function over the required period without degrading in a harmful manner.
- **Mechanical compatibility:** Should possess mechanical properties that match or complement the tissue being replaced or supported.
- **Biodegradable (for certain applications):** Some biocompatible materials are designed to degrade safely within the body over time, such as in temporary implants or drug delivery systems.

Types and examples of biocompatible materials

1. Metals
Metals are used in load-bearing implants and devices due to their strength and durability.
- **Titanium and titanium alloys:** Known for excellent biocompatibility, corrosion resistance, and mechanical properties. Used in orthopedic implants, dental implants, and cardiovascular devices.
- **Stainless steel:** Commonly used in surgical instruments and temporary implants. While less biocompatible than titanium, it is still widely used due to its strength and workability.
- **Cobalt–chromium alloys:** Used in joint replacements and dental prosthetics for their wear resistance and strength.

2. Polymers
Polymers are versatile materials used in a wide range of biomedical applications, from flexible implants to drug delivery systems:
- **PTFE:** Used in vascular grafts and surgical meshes. Known for its nonreactivity and low friction.
- **PE:** High-density PE is used in joint replacement components like acetabular cups.

- **PLA and PGA**: Biodegradable polymers used in sutures, drug delivery systems, and tissue engineering scaffolds.
- **Silicone**: Used in a variety of implants, such as breast implants and catheters, due to its flexibility and biocompatibility.

3. Ceramics

Ceramics are used in applications requiring hardness, wear resistance, and biocompatibility.

- **Alumina (aluminum oxide)**: Used in hip prostheses for its hardness and wear resistance.
- **Zirconia**: Known for its strength and toughness, it is used in dental implants and orthopedic implants.
- **Calcium phosphates (e.g., hydroxyapatite)**: Used in bone grafts and coatings for metallic implants to promote bone integration due to their similarity to natural bone mineral.

4. Composites

Composites combine materials to achieve desirable properties for specific biomedical applications.

- **Carbon fiber composites**: Used in prosthetics and implants where high strength and low weight are essential.
- **Hydroxyapatite-coated metals**: Improve the integration of metal implants with bone.

5. Natural materials

Natural materials are derived from biological sources and are used for their inherent biocompatibility:

- **Collagen**: Used in wound dressings, tissue engineering, and cosmetic surgery.
- **Chitosan**: Derived from chitin, it is used in wound healing and drug delivery systems.
- **Alginate**: Extracted from seaweed, it is used in wound dressings and tissue engineering.

Applications of biocompatible materials

1. **Orthopedic implants:**
   - Titanium and stainless steel for joint replacements.
   - PE and ceramic components for wear resistance.
2. **Dental implants:**
   - Titanium implants for their osseointegration properties.
   - Zirconia for its aesthetic and mechanical properties.

3. **Cardiovascular devices:**
   - PTFE and Dacron grafts for vascular surgery.
   - Metal stents and pacemaker leads.
4. **Drug delivery systems:**
   - Biodegradable polymers like PLA and PGA for controlled drug release.
   - Liposomes and hydrogels for targeted delivery.
5. **Tissue engineering:**
   - Scaffolds made from collagen, alginate, and synthetic polymers for tissue regeneration.
   - Hydroxyapatite coatings to enhance bone growth.
6. **Wound care:**
   - Collagen and alginate dressings for promoting healing and maintaining a moist wound environment.
   - Silicone gels and sheets for scar management.

1.14 Intelligent biological material

Intelligent biological materials, also known as smart biomaterials, are designed to interact with biological systems in dynamic and adaptive ways, responding to environmental stimuli to perform specific functions [1]. These materials can sense changes in their surroundings, process the information, and elicit a predetermined response. This capability makes them highly valuable in various biomedical applications, including drug delivery, tissue engineering, and diagnostics.

Characteristics of intelligent biological materials
- **Responsiveness:** They have the ability to respond to external stimuli such as temperature, pH, light, and specific biological molecules.
- **Biocompatibility:** They are nontoxic and non-immunogenic, ensuring safe interaction with biological tissues.
- **Adaptability:** They have the ability to change physical or chemical properties in response to environmental changes.
- **Reversibility:** They can return to their original state once the stimulus is removed (if required for the application).
- **Functionality integration:** They combine multiple functions such as sensing, actuation, and drug release in a single material system.

Types and examples of intelligent biological materials

1. Stimuli-responsive polymers
Stimuli-responsive polymers can change their physical or chemical properties in response to external stimuli [10].
- **Temperature-responsive polymers:**
    - **Example**: PNIPAM undergoes a reversible phase transition from hydrophilic to hydrophobic at a specific temperature. It is used in controlled drug release and cell culture applications.
- **pH-responsive polymers:**
    - **Example:** Poly(acrylic acid) swells or contracts in response to pH changes. This is used in targeted drug delivery systems to release drugs in specific parts of the body, such as the acidic environment of a tumor.
- **Light-responsive polymers:**
    - **Example**: Azobenzene-containing polymers that change the shape or properties when exposed to light. They are used in controlled drug delivery and photoresponsive coatings.

2. Hydrogels
Hydrogels are water-swollen, cross-linked polymer networks that can respond to environmental changes.
- **Thermoresponsive hydrogels:**
    - **Example:** Pluronic F127, which forms a gel at body temperature and a solution at lower temperatures. It is used in injectable drug delivery systems.
- **Glucose-responsive hydrogels:**
    - **Example:** Hydrogels containing glucose oxidase that can swell or shrink in response to glucose levels. They are used in diabetes management for controlled insulin delivery.

3. Self-healing materials
Self-healing materials can autonomously repair damage, enhancing the longevity and reliability of biomedical devices.
- **Example**: Hydrogels with embedded microcapsules containing healing agents are released upon damage, allowing the material to heal itself. They are used in tissue engineering and wound dressings.

4. Shape-memory materials
Shape-memory materials can return to a predefined shape when exposed to an external stimulus.

– **Example**: SMPs that change the shape in response to temperature changes. They are used in minimally invasive surgical devices and stents that expand at body temperature.

5. Magnetic and electric field-responsive materials

These materials respond to magnetic or electric fields, enabling remote control and actuation.
– **Magnetic nanoparticles:**
  – **Example**: Iron oxide nanoparticles used for targeted drug delivery and hyperthermia treatment of tumors where the particles generate heat when exposed to an alternating magnetic field.
– **EAPs:**
  – **Example**: PPy, which changes the shape when an electric field is applied. It is used in actuators and artificial muscles.

Applications of intelligent biological materials

1. **Targeted drug delivery:**
   – Smart polymers and hydrogels that release drugs in response to specific stimuli, such as pH changes in tumors or glucose levels in diabetic patients [6].
2. **Tissue engineering:**
   – Scaffolds made from responsive hydrogels that provide the right environment for cell growth and can release growth factors in response to environmental cues.
3. **Diagnostics and biosensors:**
   – Materials that change properties in the presence of specific biomarkers, providing a detectable signal for the diagnosis of diseases.
4. **Wound healing:**
   – Self-healing hydrogels that can repair themselves and provide a moist environment for optimal wound healing.
5. **Minimally invasive surgery:**
   – Shape-memory materials that can be inserted in a compact form and then expand or change the shape within the body.
6. **Wearable and implantable devices:**
   – Smart materials that can monitor physiological parameters and provide real-time feedback or therapeutic actions.

1.15 Biomimetic materials

Biomimetic materials are engineered materials that mimic the structures, functions, or processes found in biological systems. The field of biomimetics involves studying nature's designs and mechanisms to create new materials and technologies that solve human problems with improved efficiency and sustainability. These materials can replicate the sophisticated capabilities of natural substances, such as self-healing, adaptability, and environmental responsiveness.

Characteristics of biomimetic materials
- **Mimicry of biological structures**: Imitation of the hierarchical organization and complex structures found in natural materials.
- **Functional emulation:** Replication of biological functions such as self-cleaning, self-healing, or responsive behavior.
- **Material efficiency:** Use of materials in a way that maximizes performance while minimizing resource use, similar to natural systems.
- **Environmental adaptability:** Ability to interact with and adapt to the surrounding environment in a beneficial manner.

Types and examples of biomimetic materials

1. Self-healing materials
These materials can repair themselves after damage, similar to how living tissues heal.
- **Polymers with embedded microcapsules**: Containing healing agents that are released upon cracking, initiating a repair process.
    - **Example**: Polymers with microcapsules containing dicyclopentadiene, which polymerizes upon exposure to a catalyst released from a damaged area.
- **Biological inspiration**: Mimics the self-healing process of skin and bone.

2. Superhydrophobic surfaces
Materials designed to repel water, inspired by natural surfaces like lotus leaves.
- **Nanotextured surfaces**: Creating a rough surface at the nanoscale that reduces the contact area between the water and the surface.
    - **Example**: Coatings inspired by the lotus leaf effect, used in self-cleaning windows and water-repellent fabrics.
- **Biological inspiration**: Lotus leaves, which have microscopic structures that cause water droplets to bead up and roll off.

3. Biomimetic adhesives

Biomimetic adhesives are the ones that mimic the attachment mechanisms of natural organisms:

- **Gecko-inspired adhesives**: Using microscale and nanoscale structures to create strong, reversible adhesion without the use of liquids or chemical bonding.
    - **Example**: Synthetic adhesives that replicate the micro-hairs (setae) found on gecko feet, enabling strong adhesion to various surfaces.
- **Biological inspiration**: Gecko feet, which allow geckos to climb smooth vertical surfaces.

4. Biomimetic textiles

Textiles are designed to replicate the properties of natural fibers and surfaces.

- **Spider silk-inspired fibers**: Creating synthetic fibers that match the strength and elasticity of spider silk.
    - **Example**: Artificial spider silk made from recombinant DNA technology, used in high-strength, lightweight fabrics.
- **Biological inspiration**: Spider silk, which combines high tensile strength with elasticity.

5. Biomimetic structural materials

Materials that replicate the strength, flexibility, and lightweight properties of natural structures.

- **Bone-inspired composites**: Combining different materials to achieve a balance of strength and flexibility similar to natural bone.
    - **Example**: Composite materials used in lightweight, high-strength construction materials.
- **Biological inspiration**: Bone, which has a hierarchical structure that provides strength and lightness.

6. Biomimetic sensors

Biomimetic sensors are the ones that replicate the sensory functions of biological organisms.

- **Olfactory sensors**: Mimicking the sense of smell to detect specific chemical compounds.
    - **Example**: Electronic noses that use sensor arrays and pattern recognition algorithms to identify odors.
- **Biological inspiration**: Animal olfactory systems, which can detect a wide range of chemical substances with high sensitivity and specificity.

Applications of biomimetic materials
- **Medical devices and implants:**
  - Self-healing materials for long-lasting implants and prosthetics.
  - Biomimetic scaffolds for tissue engineering that replicate the extracellular matrix.
- **Environmental engineering:**
  - Superhydrophobic surfaces for self-cleaning buildings and infrastructure.
  - Water collection systems inspired by the fog-harvesting capabilities of beetles and plants.
- **Robotics and artificial intelligence:**
  - Gecko-inspired adhesives for climbing robots.
  - Biomimetic sensors for enhanced environmental perception in robots.
- **Textiles and wearables:**
  - Spider silk-inspired fibers for durable and lightweight clothing.
  - Biomimetic fabrics with enhanced breathability and water resistance.
- **Aerospace and automotive:**
  - Lightweight, high-strength materials inspired by bone and shell structures for improved fuel efficiency.
  - Self-healing coatings for aircraft and automotive surfaces.
- **Consumer products:**
  - Superhydrophobic coatings for stain-resistant clothing and self-cleaning surfaces.
  - Biomimetic adhesives for household and industrial applications.

1.16 Technological applications of intelligent materials

Intelligent materials, also known as smart materials, have the ability to respond to external stimuli such as temperature, pressure, moisture, electric or magnetic fields, and pH changes. Their unique properties make them suitable for a wide range of innovative applications across various industries. Here are some of the key technological applications of intelligent materials:

1. **Medical and healthcare**
   - **Drug delivery systems:** Smart polymers and hydrogels that release drugs in response to specific stimuli such as pH or temperature changes. For example, targeted cancer therapies, where the drug is released specifically at the tumor site.
   - **Medical implants:** SMAs and SMPs used in stents that expand at body temperature or other implantable devices that adapt to the body's conditions.
   - **Tissue engineering:** Scaffolds made from smart materials that provide the necessary environment for cell growth and can release growth factors in response to biological signals.

- **Wound healing:** Self-healing hydrogels that create a moist environment and can release antimicrobial agents in response to the presence of bacteria.

2. **Aerospace and defense**
   - **Adaptive airframes:** SMAs used in morphing wing structures that can change the shape during flight to improve aerodynamics and fuel efficiency.
   - **Thermal protection systems:** Ablative materials that protect spacecraft during reentry by absorbing and dissipating heat.
   - **Smart sensors:** Integrated sensor networks that monitor structural integrity and detect damage or stress in real time, improving maintenance and safety.

3. **Consumer electronics**
   - **Flexible electronics:** Conductive polymers and shape-memory materials are used in flexible displays, wearable devices, and foldable smartphones.
   - **Responsive wearables:** Textiles embedded with smart materials that can monitor vital signs, change color based on temperature, or provide haptic feedback.
   - **Self-healing materials:** Polymers that can repair scratches and damage on electronic devices like smartphones and tablets.

4. **Construction and Infrastructure**
   - **Smart concrete:** Concrete embedded with self-healing materials that can repair cracks autonomously, extending the lifespan of structures.
   - **Adaptive windows:** Electrochromic materials used in smart windows that can change their transparency in response to electric signals, improving energy efficiency in buildings.
   - **Structural health monitoring:** Sensor networks using piezoelectric materials to detect and report structural damage or stress in real-time.

5. **Automotive**
   - **Adaptive materials:** SMAs used in adaptive car parts such as grilles and spoilers that can change the shape to improve aerodynamics or cooling.
   - **Self-healing paints:** Polymers that can repair minor scratches and damages to maintain the aesthetic appearance of vehicles.
   - **Smart tires:** Tires embedded with sensors made from piezoelectric materials that can monitor pressure, temperature, and wear in real time.

6. **Energy**
   - **Smart batteries:** Materials that can optimize energy storage and release based on usage patterns and environmental conditions, improving battery life and efficiency.
   - **Energy harvesting:** Piezoelectric materials used to convert mechanical energy from vibrations, movements, and pressure changes into electrical energy, powering small devices or sensors.
   - **Thermoelectric materials:** Materials that convert temperature differences directly into electricity, used in power generation and cooling systems.

7. **Environmental monitoring and protection**
   - **Pollution control:** Smart materials that can detect and neutralize pollutants in air and water, such as responsive catalysts that activate in the presence of specific contaminants.
   - **Environmental sensors:** Sensors made from smart materials that monitor environmental conditions such as temperature, humidity, and pollutant levels, providing real-time data for climate studies and pollution control.

1.17 Summary

This chapter provides a comprehensive overview of different classes of materials, including their uses and advancements in intelligent materials. It covers traditional materials like metals, ceramics, polymers, and composites, emphasizing their unique properties and applications. The chapter also explores intelligent or smart materials that can dynamically respond to external stimuli such as temperature, pressure, and pH, with examples including SMAs and piezoelectric materials. It evaluates the state of materials science, discussing structural materials, functional materials, and polyfunctional materials. Additionally, it delves into the generation and diverse applications of smart materials in fields such as medicine, aerospace, and environmental monitoring, and examines biomimetic materials and their technological impacts.

Review questions

1. What are the main classes of materials and their primary characteristics?
2. How are metals, ceramics, and polymers used differently in industry?
3. What defines an intelligent or a smart material?
4. Can you provide examples of smart materials and describe their applications?
5. What are the key factors considered in the evaluation of materials science?
6. How has the field of materials science evolved over the past decades?
7. What are structural materials and why are they important in construction and engineering?
8. Describe the properties and uses of steel and concrete as structural materials.
9. What are functional materials and how do they differ from structural materials?
10. Provide examples of functional materials and their applications in technology.
11. What are polyfunctional materials?
12. How do polyfunctional materials enhance the performance of products in various industries?
13. What are the key processes involved in the generation of smart materials?
14. How do advancements in nanotechnology contribute to the development of smart materials?

15. In what diverse areas can intelligent materials be applied?
16. Discuss the role of intelligent materials in the automotive and aerospace industries.
17. What are the primitive functions of intelligent materials?
18. How do these primitive functions enable smart materials to perform specific tasks?
19. What does it mean for a material to have intelligent properties inherently?
20. Provide examples of materials with inherent intelligent properties and their uses.
21. Provide examples of traditional and advanced materials and their typical applications.
22. How do advanced materials differ from traditional materials in terms of performance?
23. What are the main types of intelligent materials?
24. How are intelligent materials revolutionizing the field of medical devices?
25. What properties make a material suitable for structural applications?
26. How do composite materials improve structural performance compared to traditional materials?
27. What are electrical materials and how are they categorized?
28. Discuss the applications of conductive polymers in electronics.
29. What are biocompatible materials and why are they crucial in the biomedical field?
30. Give examples of biocompatible materials used in implants and prosthetics.
31. Define intelligent biological materials and their primary functions.
32. How are intelligent biological materials used in drug delivery systems?
33. What is biomimetics and how does it inspire material design?
34. Provide examples of biomimetic materials and their real-world applications.
35. Discuss the technological applications of intelligent materials in environmental monitoring.
36. How are smart materials used in the development of responsive wearables?
37. What criteria are used to evaluate the performance of smart materials?
38. How do smart materials compare with traditional materials in terms of durability and efficiency?
39. What are some potential future trends in the development of intelligent materials?
40. How might advancements in intelligent materials impact future technological innovations?

2 Smart materials and structural systems

2.1 Introduction

Smart materials and structural systems [17] represent a ground-breaking shift in the design and functionality of various structures and devices. Smart materials, such as shape-memory alloys (SMAs), piezoelectric materials, and electroactive polymers (EAPs), have unique properties that enable them to change their shape, stiffness, or other characteristics in response to specific external stimuli. These materials can sense changes in their environment and react accordingly, making them ideal for a wide range of applications.

In structural systems, the integration of smart materials allows for the development of adaptive and responsive structures. For example, buildings equipped with smart materials can automatically adjust to environmental changes, such as altering their thermal properties to maintain optimal indoor temperatures, enhancing energy efficiency and occupant comfort. In the aerospace industry, smart materials can be used to create aircraft components that adjust their shape during flight to improve aerodynamic performance and fuel efficiency [13].

The medical field also benefits significantly from smart materials. Implants and prosthetics made from these materials can respond to changes in the body, providing better integration and functionality. For instance, smart stents can expand or contract in response to blood flow, ensuring optimal performance and reducing the risk of complications.

Overall, smart materials and structural systems offer a new level of innovation and adaptability, leading to smarter, more efficient, and more sustainable solutions in a variety of industries. Their ability to respond to environmental changes in real time opens up endless possibilities for the future of technology and design.

2.2 The principal ingredients of smart materials

The principal ingredients of smart materials include:
1. **SMAs:** These materials can return to their original shape after deformation when exposed to certain temperatures. They are used in applications requiring precise control and movement.
2. **Piezoelectric materials:** These materials generate an electric charge in response to mechanical stress and can also change the shape when an electric field is applied. They are used in sensors, actuators, and energy-harvesting devices [13].
3. **EAPs:** These polymers change the shape or size when stimulated by an electric field. They are used in flexible electronics, artificial muscles, and responsive surfaces [7].

https://doi.org/10.1515/9783111379623-002

4. **Magnetostrictive materials:** These materials change the shape or dimension in the presence of a magnetic field. They are used in actuators, sensors, and vibration control systems.
5. **Thermochromic and photochromic materials:** These materials change the color in response to temperature changes or light exposure. They are used in smart windows, clothing, and signage.
6. **Hydrogels:** These are polymer networks that can absorb large amounts of water and change their volume in response to environmental conditions such as pH, temperature, or electric fields. They are used in drug delivery systems, tissue engineering, and soft robotics.
7. **Chromogenic materials:** Chromogenic materials can detect and respond to environmental changes, making them ideal for smart packaging applications.
8. **Active smart polymers:** Active smart polymers possess excellent conductivity, mechanical strength, and chemical resistance, making them ideal for smart sensors.
9. **Ionic polymer–metal composites (IPMCs):** IPMCs bend in response to low voltage and generate voltage when bent, making them ideal as transducers.
10. **Electrorheological (ER) fluids:** ER fluids can provide variable forces to control mechanical vibrations, making them ideal for smart structures and machines.

2.3 Thermal materials

Thermal materials are designed to manage heat by conducting it away, insulating against it, or storing it for later release [14]. They are crucial in various applications, from electronics to building construction.

2.3.1 Types of thermal materials:

2.3.1.1 Conductive materials

Conductive materials are substances that allow the easy flow of heat or electricity. They are essential in various technological and industrial applications because they can transfer energy efficiently.

Conductive materials are characterized by their high thermal and electrical conductivity. High thermal conductivity enables these materials to transfer heat quickly, making them ideal for applications such as heat sinks in electronics, where efficient heat dissipation is crucial. High electrical conductivity allows these materials to carry electric current efficiently, so metals like copper are widely used in electrical wiring and circuitry. Figure 2.1 illustrates how conductive materials enable the flow of both heat and electricity, highlighting their high thermal and electrical conductivities. On one side, the diagram shows heat transfer through a material with arrows representing the flow of heat, making it ideal for applications such as heat sinks in electronics, where rapid heat dissipation is critical. The other side of the figure focuses on electrical conductivity, depicting **copper wiring** efficiently carrying electric current, represented by arrows indicating the flow of electricity. This showcases how conductive materials are essential in both thermal management and electrical systems due to their ability to transfer energy efficiently. Common examples of conductive materials include metals such as copper, aluminum, and gold. Copper is renowned for both its thermal and electrical conductivities, aluminum is valued for its lightweight and good conductivity, and gold, while less common due to its cost, is used in specialized applications for its excellent conductivity and resistance to corrosion.

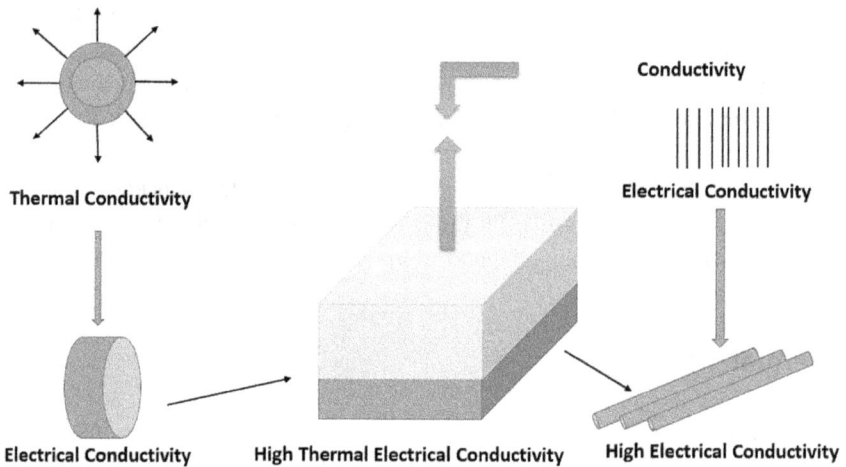

Figure 2.1: Heat and electrical flow in conductive materials.

Applications:
– **Heat sinks**:
  – **Material**: Aluminum or copper.
  – **Use**: In computers, to draw heat away from processors, ensuring they stay cool and perform efficiently.
– **Electrical wiring**:
  – **Material**: Copper.
  – **Use**: Used in homes and electronic devices for its excellent electrical conductivity, ensuring efficient transmission of electricity with minimal loss.
– **Cooking utensils**:
  – **Material**: Stainless steel or aluminum.
  – **Use**: Pots and pans made from these materials distribute heat evenly, improving the cooking efficiency.

Practical example: conductive materials

Copper in electrical wiring
Copper is a prime example of a conductive material widely used in electrical wiring due to its excellent electrical conductivity, ductility, and relatively low cost.

– **Scenario:**
In a new building, the electrical system needs to be installed to ensure a reliable power supply to all outlets and devices.

– **Solution:**
Copper wires are chosen for the electrical circuits.

– **Function:**
Copper's high electrical conductivity allows it to efficiently transmit electricity with minimal resistance and energy loss. This ensures that electrical power is distributed effectively throughout the building.

– **Outcome:**
The use of copper wiring results in a reliable and efficient electrical system. The building experiences minimal energy loss, reducing electricity costs and increasing safety by preventing overheating and potential fire hazards. Additionally, the flexibility and durability of copper make the installation process easier and the wiring system more robust.

By using copper wires, the building's electrical system benefits from enhanced performance, safety, and cost-effectiveness, demonstrating the practical advantages of conductive materials in everyday applications.

2.3.1.2 Insulating materials
Insulating materials are designed to resist the flow of heat and electricity, making them essential for various applications where controlling energy transfer is crucial. These materials have low thermal and electrical conductivity, helping to maintain temperatures and prevent electrical currents from passing through. Their low thermal conductivity allows them to resist the heat flow, keeping environments stable, while their low electrical conductivity prevents electrical currents from passing through, ensuring safety in electrical applications. Common examples of insulating materials include carbon fiber, foam, rubber, and ceramics. These materials are widely used in building insulation, electrical insulation, and thermal insulation in appliances.

Applications
– **Building insulation**:
    – **Material**: Carbon fiber

Figure 2.2 illustrates carbon fiber insulation, one of the most thoroughly tested building materials, commonly installed as batt or blown-in insulation. Carbon fiber is eco-friendly, nonflammable, safe to install, and maintains its thermal performance for the lifetime of the building.

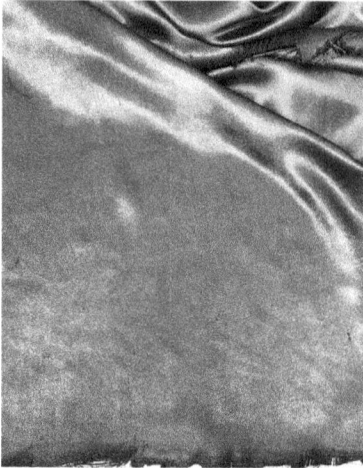

Figure 2.2: Carbon fiber in mat form.

The advantages of carbon fiber include:
- **Eco-friendly**: Carbon fiber is environmentally sustainable, and made from recycled glass spun into fibers.
- **Safe**: It is nonflammable and safe to install in homes and buildings.
- **Durable**: Carbon fiber maintains its thermal performance throughout the life of the building, offering reliable protection in both warm and cold climates.
    - **Use**: Installed in walls, attics, and floors to maintain indoor temperatures by reducing heat loss in winter and heat gain in summer.
- **Electrical insulation**:
    - **Material**: Rubber.

Figure 2.3 depicts the cross section of an electrical wire or a cable designed to prevent electrical shocks and short circuits. It typically includes an insulating outer layer made of materials like PVC or rubber to prevent current leakage and protect users from shocks. Beneath this insulation is the conductor, usually copper or aluminum, which carries the electrical current. Additional protective layers may be present to shield against physical damage and environmental factors. In some cases, a shielding layer is also included to reduce electromagnetic interference and maintain signal integrity. Together, these layers ensure safe and reliable operation of electrical wires and cables.

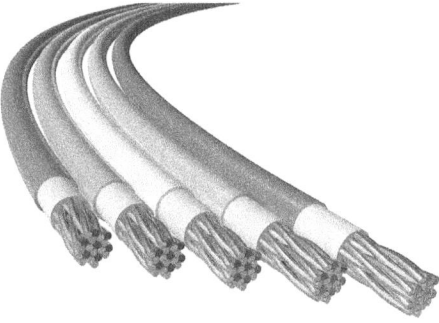

Figure 2.3: Rubber (https://expressassembliesltd blog.wordpress.com/wp-content/uploads/2014/ 02/wire-and-cable.jpg) (accessed on September 18, 2024).

- **Use**: Coating for electrical wires and cables to prevent electrical shocks and short circuits.
- **Thermal insulation in appliances**:
  - **Material**: Foam.
  - **Use**: Used in refrigerators and freezers to maintain low temperatures by minimizing heat exchange with the environment.

Practical example: carbon fiber insulation in homes
- **Scenario**: Insulating a house to improve energy efficiency.
- **Material**: Carbon fiber batts are installed in the walls and attic.
- **Function**: The carbon fiber traps air, reducing the flow of heat between the inside and outside of the house.
- **Outcome**: The home remains warmer in winter and cooler in summer, leading to reduced energy bills and increased comfort.

2.3.1.3 Phase change materials (PCMs)

Phase change materials (PCMs) are substances that absorb and release thermal energy during the process of melting and freezing, as shown in Figure 2.4. They are used to regulate the temperature in various applications by storing and releasing heat. PCMs can change their state from solid to liquid and vice versa at specific temperatures, allowing them to store large amounts of energy in the form of latent heat. This property makes them ideal for thermal management in building materials, solar energy systems, and thermal packaging. Common examples include paraffin wax and salt hydrates, which are used to enhance energy efficiency and provide consistent temperature control in diverse settings.

Applications of PCMs
- **Building materials**:
  - **Use**: Incorporated into walls, ceilings, and floors.

- **Benefit**: PCMs absorb excess heat during the day and release it at night, maintaining a stable indoor temperature and reducing heating and cooling costs.
- **Solar energy systems**:
  - **Use**: Integrated into solar panels and thermal storage units.
  - **Benefit**: PCMs store heat from the Sun during the day and release it when sunlight is not available, providing a continuous energy supply and improving the efficiency of solar power systems.
- **Thermal packaging**:
  - **Use**: Used in packaging for temperature-sensitive products such as pharmaceuticals and food.
  - **Benefit**: PCMs maintain a consistent temperature during transportation, ensuring the integrity and quality of the products.
- **Textiles and clothing**:
  - **Use**: Incorporated into fabrics and garments.
  - **Benefit**: PCMs regulate body temperature by absorbing, storing, and releasing heat, providing comfort in varying environmental conditions.
- **Electronics cooling**:
  - **Use**: Integrated into cooling systems for electronic devices.
  - **Benefit**: PCMs absorb heat generated by electronic components, preventing overheating and enhancing the performance and lifespan of devices.
- **Refrigeration and air-conditioning**:
  - **Use**: Used in cooling systems to store and release cold energy.
  - **Benefit**: PCMs improve the efficiency of refrigeration and air-conditioning systems by maintaining stable temperatures and reducing energy consumption.

Practical example: PCM in building materials
- **Scenario**: A new office building is designed to be energy-efficient.
- **Solution**: PCMs are integrated into the building's walls and ceiling.
- **Function**: During the day, PCMs absorb heat, preventing indoor temperatures from rising too high. At night, the absorbed heat is released, warming the building as outdoor temperatures drop.
- **Outcome**: The building maintains a more consistent temperature throughout the day and night, reducing the need for additional heating and cooling, and thus lowering energy bills and improving occupant comfort.

2.4 Sensing technologies

Sensing technologies refer to a broad array of devices and systems designed to detect, measure, and respond to physical properties or environmental changes. These technologies convert real-world information into data that can be analyzed and used for

various applications. Sensors can measure a wide range of variables, including temperature, pressure, light, motion, and chemical composition.
- **Characteristics:**
  - **Detection capability**: Ability to sense specific physical or environmental changes.
  - **Signal conversion**: Transforms detected information into electrical signals or data.
  - **Accuracy and precision**: High accuracy and precision in measurements for reliable data.
- **Common examples:**
  - **Temperature sensors**: Measure temperature changes (e.g., thermocouples and resistive temperature devices).
  - **Pressure sensors**: Detect pressure variations in gases or liquids (e.g., barometers and piezoelectric sensors).
  - **Light sensors**: Measure light intensity (e.g., photodiodes and light-dependent resistors).
  - **Motion sensors**: Detect movement (e.g., accelerometers and gyroscopes).
  - **Chemical sensors**: Identify chemical compositions and concentrations (e.g., pH sensors and gas detectors).
- **Applications:**
  - **Healthcare**: Monitoring vital signs with sensors like heart rate monitors and glucose sensors.
  - **Industrial automation**: Using pressure and temperature sensors to control manufacturing processes.
  - **Consumer electronics**: Implementing light and motion sensors in smartphones for screen adjustment and activity tracking.
  - **Environmental monitoring**: Employing chemical sensors to detect pollutants and monitor air quality.
  - **Smart homes**: Integrating various sensors for home automation, such as temperature sensors for climate control and motion sensors for security systems.

Practical example: temperature sensors in heating, ventilation, and air-conditioning (HVAC) systems
- **Scenario**: Maintaining optimal indoor climate in a building.
- **Solution**: Temperature sensors are installed within the HVAC system.
- **Function**: The sensors continuously monitor the indoor temperature and relay the data to the HVAC control unit.
- **Outcome**: The HVAC system adjusts heating or cooling based on the sensor data, ensuring a comfortable indoor environment while optimizing energy efficiency.

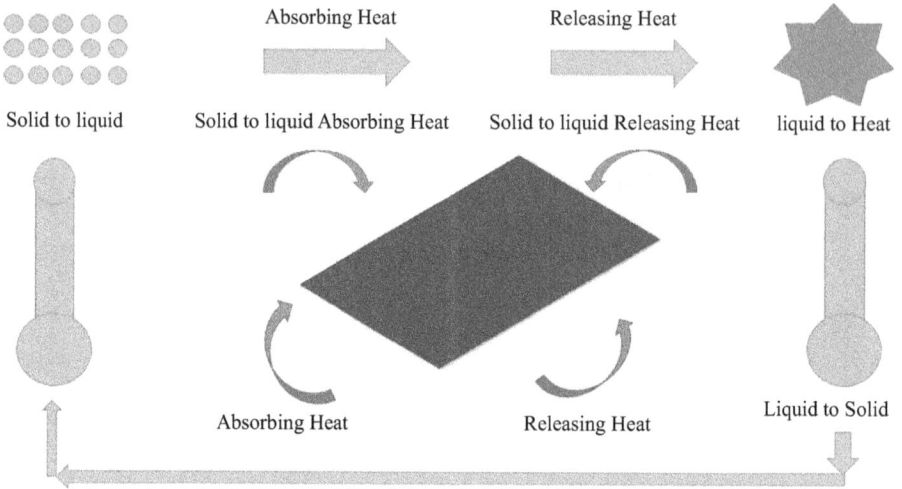

Absorbing Heat

Releasing Heat

Solid to liquid

Solid to liquid Absorbing Heat

Solid to liquid Releasing Heat

liquid to Heat

Absorbing Heat

Releasing Heat

Liquid to Solid

Figure 2.4: Thermal energy absorption and release in phase change materials (PCMs).

2.5 Microsensors

Microsensors are miniature sensors designed to detect and measure physical, chemical, or biological changes at a very small scale. Due to their small size, they can be integrated into various applications where space is limited, and high precision is required.

- **Characteristics:**
  - **Miniature size**: Extremely small dimensions allow for integration into compact systems.
  - **High sensitivity**: Capable of detecting minute changes in the environment.
  - **Low power consumption**: Designed to operate efficiently with minimal energy use.
- **Common examples:**
  - **MEMS (microelectromechanical systems) accelerometers**: Measure acceleration and tilt, commonly used in smartphones and wearable devices.
  - **Micropressure sensors**: Detect pressure changes in medical devices and industrial applications.
  - **Microtemperature sensors**: Monitor temperature in electronic circuits and biomedical applications.
  - **Microchemical sensors**: Detect specific chemicals or gases used in environmental monitoring and medical diagnostics.
- **Applications:**
  - **Consumer electronics**: Integrated into smartphones, tablets, and wearables for functions like screen rotation, step counting, and health monitoring.

- **Medical devices**: Used in minimally invasive medical equipment to monitor vital signs and diagnose conditions.
- **Automotive industry**: Implemented in vehicles for airbag deployment systems, tire pressure monitoring, and stability control.
- **Industrial automation**: Used in machinery and robotics for precise control and monitoring of processes.
- **Environmental monitoring**: Employed in detecting pollutants and monitoring air and water quality on a microscale.

Practical example: MEMS accelerometers in smartphones
- **Scenario**: Enabling screen rotation in smartphones.
- **Solution**: MEMS accelerometers are embedded in the device.
- **Function**: The accelerometer senses the orientation of the phone by measuring the direction of gravitational pull.
- **Outcome**: When the user rotates the phone, the screen orientation changes automatically to match the new position, enhancing user experience.

Microsensors provide precise measurements and enable advanced functionalities in compact and portable devices, making them indispensable in modern technology.

2.6 Hybrid smart materials

Hybrid smart materials are advanced materials engineered to combine multiple properties or functionalities within a single substance. They can respond to external stimuli, such as temperature, light, pressure, or magnetic fields, by changing their properties or behavior in a controlled manner. By integrating different materials or components at the molecular or nanolevel, hybrid smart materials offer enhanced performance and new capabilities [13].
- **Characteristics:**
  - **Multifunctionality**: Ability to perform multiple functions due to the integration of different material properties.
  - **Responsiveness**: Ability to respond to various external stimuli in a predictable and reversible manner.
  - **Adaptability**: Capable of adapting their properties based on environmental changes.
- **Common examples:**
  - **Piezoelectric polymers**: Generate electrical charge in response to mechanical stress and vice versa.
  - **SMAs coated with polymers**: Combine the shape memory effect of metals with the flexibility and corrosion resistance of polymers.

  - **Conductive polymers with embedded nanoparticles**: Enhance electrical conductivity and add functional properties such as sensing or catalysis.
- **Applications:**
  - **Medical devices**: Used in stents, prosthetics, and drug delivery systems that adapt to body conditions.
  - **Aerospace**: Materials that adjust to temperature changes, pressure, or vibrations to enhance performance and durability.
  - **Consumer electronics**: Flexible displays and wearable devices that require materials to change shape or conductivity.
  - **Construction**: Smart concrete that can self-heal cracks or adapt to environmental conditions to improve longevity.
  - **Textiles**: Fabrics that change the color, regulate the temperature, or monitor health parameters.

Practical example: SMAs in medical stents
- **Scenario**: Developing a stent that can be easily inserted into a blocked artery and then expand to hold it open.
- **Solution**: SMAs are used in the stent design.
- **Function**: The stent is compacted for insertion. Once in place, body temperature triggers the SMA to expand to its preset shape, securing the artery.
- **Outcome**: The stent effectively holds the artery open, improving blood flow and reducing the risk of complications, while its adaptability ensures a minimally invasive procedure.

Hybrid smart materials are revolutionizing various fields by providing innovative solutions that leverage their multifunctional and responsive properties to enhance performance and functionality.

2.7 Algorithm for synthesizing smart material

Synthesizing a smart material involves a systematic algorithmic approach to combine different components and properties to achieve the desired responsive behaviors [9]. Here is a brief outline of the typical algorithm for synthesizing a smart material:

Algorithm for synthesizing a smart material
- **Define objectives and requirements:**
  - **Objective**: Determine the desired functionalities and applications (e.g., self-healing, shape memory, and conductivity).
  - **Requirements**: Establish specific requirements such as mechanical strength, temperature range, response time, and biocompatibility.

– **Select base materials:**
  – **Identification**: Choose base materials that inherently possess some of the desired properties (e.g., polymers, metals, and ceramics).
  – **Compatibility**: Ensure compatibility between selected materials for effective integration.
– **Incorporate functional components:**
  – **Additives**: Identify and incorporate functional additives or fillers (e.g., nanoparticles, SMAs, and conductive fibers) to impart additional properties.
  – **Proportions**: Determine the optimal proportions of each component to balance the properties.
– **Design the material structure:**
  – **Microstructure**: Design the microstructure or molecular architecture to achieve the desired behavior (e.g., layering, embedding nanoparticles, creating composites).
  – **Simulation**: Use computational modeling to simulate and predict the material's behavior and optimize the design.
– **Synthesis process:**
  – **Method selection**: Choose an appropriate synthesis method (e.g., sol–gel process, electrospinning, 3D printing, and chemical vapor deposition).
  – **Procedure**: Follow a precise procedure to combine and process the materials, ensuring homogeneity and stability.
– **Characterization and testing:**
  – **Physical and chemical properties**: Characterize the synthesized material using techniques like X-ray diffraction, scanning electron microscopy (SEM), and spectroscopy.
  – **Performance testing**: Conduct performance tests to evaluate the material's response to external stimuli (e.g., stress, temperature, and electrical field).
– **Optimization:**
  – **Iteration**: Iterate the synthesis process by tweaking material composition, structure, and processing conditions to optimize performance.
  – **Feedback loop**: Use test results to refine the design and synthesis parameters.
– **Scale-up and application:**
  – **Scaling**: Develop scalable synthesis methods for large-scale production.
  – **Integration**: Integrate the smart material into the intended application or device, ensuring compatibility and functionality.

Practical example: developing a self-healing polymer
– **Objective**: Create a polymer that can autonomously repair cracks.
– **Base material**: Select a flexible polymer with good mechanical properties.
– **Functional component**: Incorporate microcapsules filled with a healing agent (e.g., resin).

- **Structure design**: Disperse microcapsules uniformly within the polymer matrix.
- **Synthesis method**: Use melt blending to combine the polymer and microcapsules.
- **Characterization**: Analyze the distribution of microcapsules using SEM.
- **Testing**: Induce cracks and observe the self-healing process under controlled conditions.
- **Optimization**: Adjust the concentration and type of healing agent for improved efficiency.
- **Scale-up**: Develop a process for producing the self-healing polymer in bulk.
- **Application**: Use the polymer in coatings or structural components where durability is critical.

This algorithmic approach ensures a systematic and efficient process for developing advanced smart materials with tailored properties for specific applications.

2.8 Passive sensory smart structure

Passive sensory smart structures are systems designed to monitor and respond to environmental changes without requiring active input or power for their sensing functions. These structures integrate passive sensors, which rely on intrinsic properties to detect changes, and can autonomously gather data on various parameters like stress, temperature, or deformation. The gathered data can then be analyzed to assess the condition and performance of the structure.
- **Characteristics:**
    - **Self-sufficient**: Operate without external power sources for sensing.
    - **Durable**: Designed to function over long periods with minimal maintenance.
    - **Embedded sensors**: Sensors are often embedded within the structure itself.
    - **Real-time monitoring**: Provide continuous monitoring of environmental conditions or structural health.
- **Common examples:**
    - **Fiber-optic sensors**: Used to detect strain or temperature changes in bridges and buildings.
    - **Piezoelectric sensors**: Convert mechanical stress into electrical signals, useful in monitoring vibrations or pressure.
    - **Thermochromic materials**: Change color in response to temperature variations, indicating thermal conditions.
    - **SMAs**: Change the shape under certain conditions, indicating stress or deformation.

- **Applications:**
  - **Structural health monitoring**: Used in bridges, buildings, and other infrastructure to detect stress, strain, and potential damage, ensuring safety and longevity.
  - **Aerospace**: Monitor stress and deformation in aircraft components to prevent failure and improve maintenance schedules.
  - **Automotive industry**: Integrated into vehicle components to monitor wear and tear, enhancing safety and performance.
  - **Energy sector**: Used in wind turbines and pipelines to monitor structural integrity and detect potential failures.

Practical example: fiber-optic sensors in bridges
- **Scenario**: Monitoring the structural health of a bridge to ensure safety.
- **Solution**: Fiber-optic sensors are embedded within the bridge's concrete and steel components.
- **Function**: These sensors detect strain and temperature changes by measuring variations in light transmission through the optical fibers.
- **Outcome**: Continuous real-time data on the bridge's structural integrity is provided, enabling early detection of potential issues such as cracks or excessive stress. This allows for timely maintenance and repairs, ensuring the bridge remains safe and functional.

2.9 Reactive actuator-based smart structures

Reactive actuator-based smart structures are advanced systems that not only sense environmental changes but also actively respond to them using actuators. These structures integrate sensors to detect stimuli and actuators to perform actions, enabling them to adapt dynamically to the changing conditions. This active response capability makes them suitable for applications requiring immediate and precise adjustments.
- **Characteristics:**
  - **Active response**: Ability to perform actions in response to detected changes.
  - **Integrated sensors and actuators**: Combination of sensing and actuation components within the structure.
  - **Adaptive**: Can adjust their properties or behavior in real time.
  - **Precision control**: High level of control over responses to specific stimuli.
- **Common examples:**
  - **Piezoelectric actuators**: Convert electrical signals into mechanical movement, used in vibration control and precision positioning.
  - **SMAs**: Change the shape in response to temperature changes, used in adaptive components and medical devices.

- **EAPs**: Change the shape or size when electrically stimulated, used in soft robotics and adaptive optics.
  - **Hydraulic or pneumatic actuators**: Provide force or movement using fluid pressure, used in heavy machinery and adaptive structural elements.
- **Applications:**
  - **Aerospace**: Adaptive wings and control surfaces that adjust for optimal performance and fuel efficiency.
  - **Robotics**: Robots with actuators that allow for precise movement and manipulation, enhancing functionality and adaptability.
  - **Medical devices**: Implants and prosthetics that adapt to physiological changes, improving patient outcomes.
  - **Civil engineering**: Buildings and bridges with actuators that mitigate vibrations and seismic activities, enhancing structural safety.
  - **Automotive**: Adaptive suspension systems that adjust to road conditions for improved comfort and handling.

Practical example: adaptive aircraft wings
- **Scenario**: Improving the performance and fuel efficiency of an aircraft.
- **Solution**: Aircraft wings equipped with SMAs as actuators.
- **Function**: SMAs change their shape in response to temperature changes controlled by the aircraft's systems, altering the wing's shape to optimize aerodynamics during different flight phases.
- **Outcome**: The adaptive wings adjust in real time to various flight conditions, reducing drag, improving fuel efficiency, and enhancing overall performance and stability.

2.10 Active sensing and reactive smart structures

Active sensing and reactive smart structures are advanced systems that combine the capabilities of sensing environmental changes and actively responding to them. These structures integrate sensors to detect stimuli and actuators to perform actions, allowing them to adapt dynamically to the changing conditions [15]. The active sensing component involves continuous monitoring and data collection, while the reactive component enables immediate and precise responses.
- **Characteristics:**
  - **Active sensing**: Continuous monitoring and data collection from the environment.
  - **Reactive response**: Ability to perform actions based on sensed data.
  - **Integrated systems**: Combination of sensors and actuators within the same structure.
  - **Adaptability**: Real-time adjustments to changing conditions.
  - **Precision**: High level of control over sensing and responses.

- **Common examples:**
  - **Piezoelectric systems**: Use piezoelectric sensors and actuators to detect and mitigate vibrations in structures.
  - **SMAs**: Sensors detect temperature changes, and SMAs adjust the shape accordingly.
  - **EAPs**: Sensors detect electrical signals, and EAPs change the shape or size in response.
  - **Fiber-optic systems**: Sensors measure the strain or temperature, and actuators adjust the structure to maintain integrity.
- **Applications:**
  - **Aerospace**: Aircraft with adaptive wings and control surfaces that respond to aerodynamic changes for optimal performance.
  - **Robotics**: Robots with sensors and actuators that enable precise movements and adaptation to various tasks and environments.
  - **Medical devices**: Implants and prosthetics that adjust to physiological changes for better patient outcomes.
  - **Civil engineering**: Buildings and bridges with systems that detect and respond to seismic activities or structural stress, enhancing safety.
  - **Automotive**: Vehicles with adaptive systems that adjust suspension and other components for improved comfort and handling.

Practical example: adaptive vibration control in buildings
- **Scenario**: Mitigating the effects of vibrations and earthquakes in high-rise buildings.
- **Solution**: Integration of piezoelectric sensors and actuators within the building structure.
- **Function**: Sensors continuously monitor vibrations and structural stress. When abnormal vibrations are detected, actuators generate counter-vibrations to cancel them out.
- **Outcome**: The building maintains stability and structural integrity during earthquakes or strong winds, enhancing safety for occupants.

2.11 Smart skins: aeroelastic tailoring of airfoils

Smart skins with aeroelastic tailoring of airfoils refer to advanced materials and structural designs integrated into aircraft wings and control surfaces to optimize their aerodynamic performance [16]. These smart skins can adapt their shape and properties in response to aerodynamic forces, enhancing efficiency, reducing drag, and improving control.
- **Characteristics:**
  - **Aeroelastic tailoring**: Customizing the stiffness and flexibility of the airfoil to optimize aerodynamic performance.

- **Smart materials**: Use of materials that can change their properties in response to external stimuli (e.g., temperature and pressure).
- **Adaptive shape**: Ability to adjust the shape of the airfoil in real time based on flight conditions.
- **Enhanced performance**: Improved lift-to-drag ratio, fuel efficiency, and overall aerodynamic efficiency.
- **Components:**
  - **SMAs**: Embedded in the skin to change the shape in response to temperature changes.
  - **Piezoelectric actuators**: Generate movements or shape changes when electrical voltage is applied.
  - **Flexible composites**: Materials that can bend and flex while maintaining structural integrity.
  - **Sensors**: Integrated to monitor aerodynamic forces and structural stresses.
- **Applications:**
  - **Commercial aviation**: Enhancing the performance and fuel efficiency of passenger aircraft.
  - **Military aircraft**: Improving maneuverability and stealth capabilities by dynamically changing the airfoil shape.
  - **Unmanned aerial vehicles (UAVs)**: Optimizing flight performance for drones and other UAVs.
  - **Renewable energy**: Used in wind turbine blades to adjust to changing wind conditions for better efficiency.

Practical example: adaptive wing in commercial aircraft
- **Scenario**: Improving fuel efficiency and performance of a commercial aircraft.
- **Solution**: Integrating smart skins with aeroelastic tailoring into the aircraft's wings.
- **Function**: During different phases of flight, sensors detect aerodynamic forces. The smart skin adjusts the wing's shape using embedded SMAs and piezoelectric actuators to optimize the lift-to-drag ratio.
- **Outcome**: The aircraft experiences reduced drag and improved fuel efficiency, leading to lower operating costs and enhanced performance across various flight conditions.

2.12 Synthesis of future smart systems

The synthesis of future smart systems involves the integration of advanced technologies, materials, and algorithms to create systems that can autonomously sense, process, and react to their environment [3, 18]. These systems are designed to be highly adaptive, efficient, and intelligent, capable of operating in complex and dynamic environments.

- **Key elements:**
  - **Advanced materials**: Use of smart materials such as SMAs, piezoelectric materials, and nanomaterials that can respond to stimuli (e.g., temperature, pressure, and electric fields).
  - **Artificial intelligence (AI)**: Integration of AI and machine learning algorithms to enable systems to learn from data, make decisions, and improve performance over time.
  - **Sensor networks**: Deployment of distributed sensors to gather real-time data on environmental conditions, structural health, and system performance.
  - **Actuation mechanisms**: Implementation of actuators that can perform precise actions in response to sensor inputs, enabling adaptive responses.
  - **Energy efficiency**: Focus on energy-harvesting and low-power designs to ensure sustainable operation, especially in remote or autonomous applications.
  - **Connectivity**: Use of Internet of things (IoT) technology to enable communication between smart systems and other devices or networks [7].
- **Characteristics:**
  - **Autonomous operation**: Ability to function independently with minimal human intervention.
  - **Real-time adaptation**: Continuous monitoring and immediate response to environmental changes or system demands.
  - **Interoperability**: Seamless integration with other systems and technologies.
  - **Scalability**: Designed to be scalable for various applications, from small devices to large infrastructures.
  - **Resilience**: Built with redundancy and self-healing capabilities to ensure reliability in diverse conditions.
- **Applications:**
  - **Smart cities**: Urban infrastructure that can autonomously manage resources, traffic, energy, and waste, improving sustainability and quality of life.
  - **Healthcare**: Advanced medical devices and systems that provide personalized care, monitor health, and deliver treatments autonomously.
  - **Agriculture**: Precision farming systems that adapt to changing weather, soil conditions, and crop needs, optimizing yield and resource use.
  - **Transportation**: Autonomous vehicles and smart transportation networks that enhance safety, efficiency, and reduce the environmental impact.
  - **Industrial automation**: Factories with intelligent machinery that can self-optimize, diagnose issues, and adapt to production demands.

Practical example: autonomous smart buildings
- **Scenario**: Developing buildings that manage energy, climate, and security autonomously.
- **Solution**: Synthesize a system that integrates AI, smart materials, and IoT-enabled sensors [8].
- **Function**: The building's smart system monitors occupancy, weather conditions, and energy usage. It adjusts lighting, heating, cooling, and security systems in real time to optimize energy consumption and ensure occupant comfort.
- **Outcome**: The building operates efficiently with reduced energy costs, enhanced security, and improved occupant satisfaction, all while minimizing the environmental impact.

The synthesis of future smart systems will revolutionize various sectors by creating intelligent, adaptive, and sustainable solutions that enhance performance, efficiency, and quality of life. These systems will be pivotal in addressing the challenges of the future, from urbanization to environmental sustainability [4].

2.13 Summary

This chapter explores the integration of advanced technologies and materials into intelligent structures. It covers the essential components of smart materials, including thermal materials for heat management and sensing technologies that enable real-time data collection. Microsensors and intelligent systems play a key role in enabling adaptive behaviors, while hybrid smart materials combine multiple properties for enhanced functionality. The chapter outlines an algorithm for synthesizing smart materials, detailing the process of combining base materials and functional components. It distinguishes between passive sensory smart structures, which monitor without active input, and reactive actuator-based systems, which actively respond to stimuli. Additionally, it discusses active sensing and reactive smart structures that provide dynamic adaptation, smart skins for aeroelastic tailoring of airfoils to optimize aerodynamics, and the synthesis of future smart systems integrating AI, sensor networks, and energy-efficient technologies for autonomous and adaptable applications.

Review questions

1. What are smart materials and how do they differ from conventional materials?
2. What are the primary characteristics of smart materials?
3. Name three examples of smart materials and their key properties.
4. How do smart materials respond to environmental stimuli?
5. What role do smart materials play in enhancing structural performance?

6. What is thermal conductivity, and why is it important in smart materials?
7. How do thermal materials manage heat transfer in structural systems?
8. Describe a practical application of thermal materials in aerospace.
9. What are PCMs and how do they function?
10. How do thermal materials contribute to energy efficiency in buildings?
11. What are the key components of sensing technologies in smart systems?
12. How do sensing technologies enable real-time monitoring?
13. Explain the difference between passive and active sensors.
14. Name two common types of sensors used in structural health monitoring.
15. How do sensing technologies contribute to the safety and reliability of infrastructure?
16. What are microsensors, and what advantages do they offer?
17. How do MEMS sensors work?
18. Provide an example of a microsensor used in consumer electronics.
19. What challenges are associated with the deployment of microsensors in medical devices?
20. How do microsensors contribute to precision in industrial automation?
21. What defines an intelligent system in the context of smart materials?
22. How do AI and machine learning integrate with intelligent systems?
23. Name an application of intelligent systems in healthcare.
24. Describe how intelligent systems enhance autonomous vehicles.
25. What role do intelligent systems play in smart city infrastructure?
26. What are hybrid smart materials, and how do they combine different properties?
27. Provide an example of a hybrid smart material and its application.
28. How do hybrid smart materials improve performance compared to single-function materials?
29. What challenges are involved in designing hybrid smart materials?
30. How can hybrid smart materials be used in aerospace engineering?
31. What are the key steps in the algorithm for synthesizing a smart material?
32. How do you select base materials for a smart material synthesis?
33. What is the role of computational modeling in smart material synthesis?
34. Describe the importance of material characterization in the synthesis process.
35. How do iterative testing and optimization improve smart material performance?
36. What are passive sensory smart structures and how do they function?
37. How do passive sensory systems monitor structural conditions?
38. Give an example of a passive sensory smart structure in civil engineering.
39. What are the advantages of using passive sensory systems over active systems?
40. How do passive sensory smart structures contribute to long-term infrastructure maintenance?
41. What are reactive actuator-based smart structures?
42. How do actuators in these structures respond to environmental changes?
43. Describe a practical application of reactive actuator-based smart structures.

44. What are the benefits of integrating actuators with sensors in smart systems?
45. How do reactive actuator-based structures enhance performance in aerospace applications?
46. What is the difference between active sensing and reactive smart structures?
47. How do active sensing systems contribute to real-time adaptation?
48. Provide an example of an application where active sensing and reactive systems are used together.
49. What are the key benefits of integrating active sensing with reactive actuation?
50. How do active sensing and reactive smart structures improve the overall system efficiency?

3 Electrorheological (fluids) smart materials

3.1 Introduction

This chapter introduces a unique class of materials that exhibit rapid and reversible changes in their mechanical properties under the influence of an electric field. These smart fluids, known as electrorheological (ER) fluids, consist of a suspension of fine particles in a carrier liquid [18]. When subjected to an electric field, the viscosity of these fluids can change dramatically, transitioning from a liquid to a semisolid state within milliseconds. This controllable behavior makes ER fluids particularly valuable in applications requiring precise and adaptive control of mechanical properties, such as in vibration dampers, clutches, and adaptive shock absorbers. The chapter explores the fundamental principles behind ER fluids, including their composition, the mechanism of the ER effect, and the factors influencing their performance. It also delves into the various applications of ER smart materials, highlighting their potential to revolutionize industries by providing versatile and responsive solutions for dynamic systems.

3.2 Suspensions and electrorheological fluids

Suspensions are mixtures in which fine solid particles are dispersed in a liquid, but unlike solutions, these particles do not dissolve and may settle over time. ER fluids are a specific type of suspension where the dispersed particles are typically made of polymers, ceramics, or other materials, and are suspended in an insulating oil or similar liquid. What makes ER fluids unique is their ability to change viscosity dramatically when exposed to an electric field. This change can transform the fluid from a free-flowing liquid to a gel-like or semisolid state almost instantaneously. The effect is reversible and can be precisely controlled by adjusting the strength of the electric field, making ER fluids ideal for applications that require variable resistance or damping, such as in adaptive shock absorbers, clutches, and other devices where real-time control of fluid dynamics is essential.

- **Characteristics:**
  - **Viscosity control**: ER fluids can rapidly change their viscosity from a liquid to a semisolid when exposed to an electric field.
  - **Reversible and tunable**: The change in state is reversible, and the degree of viscosity change can be precisely controlled by varying the strength of the electric field.

https://doi.org/10.1515/9783111379623-003

- **Particle suspension**: ER fluids are suspensions of fine solid particles within a nonconductive liquid, where the particles align to form chains under an electric field, increasing the fluid's resistance to flow [19].
  - **Rapid response**: The transition between states occurs almost instantaneously, making ER fluids highly responsive to external stimuli.
- **Common examples:**
  - **Silicon oil-based ER fluids**: Silicon oil with suspended polymer or ceramic particles is commonly used in ER fluid formulations.
  - **Cornstarch in oil**: A simple example, where cornstarch particles are suspended in oil, can demonstrate ER effects in a basic setup.
- **Applications:**
  - **Adaptive shock absorbers**: Used in automotive suspension systems to dynamically adjust damping properties based on road conditions.
  - **Clutches and brakes**: ER fluids allow for precise control of torque transfer, making them ideal for applications in machinery and robotics.
  - **Vibration dampers**: Employed in buildings, machinery, and other structures to absorb and mitigate vibrations by adjusting fluid properties in real time.
  - **Tactile feedback devices**: Utilized in haptic devices to provide variable resistance, enhancing the realism of virtual reality or gaming experiences.

Practical example: adaptive car suspension system

- **Scenario**: A vehicle requires adjustable suspension to maintain comfort and stability across various driving conditions.
- **Solution**: The car's suspension system incorporates ER fluid-filled dampers.
- **Function**: When the car encounters rough terrain, an electric field is applied to the ER fluid, increasing its viscosity and stiffening the suspension. On smooth roads, the electric field is reduced, allowing the suspension to become softer for a more comfortable ride.
- **Outcome**: The system continuously adapts to driving conditions, providing optimal comfort and handling, improving the overall ride quality and safety.

3.3 Bingham body model

The Bingham body model describes the behavior of materials that exhibit both solid and fluid characteristics, specifically in response to applied stress. According to this model, a material behaves like a rigid solid at low-stress levels but flows like a viscous fluid once a certain threshold of stress, known as the yield stress, is exceeded.

- **Characteristics:**
  - **Yield stress**: The minimum stress required to initiate flow. Below this stress, the material does not deform or flow.
  - **Viscous flow**: Once the yield stress is surpassed, the material flows with a constant viscosity, similar to a Newtonian fluid.
  - **Plastic behavior**: The model captures the transition from solid-like behavior to fluid-like behavior.
- **Common examples:**
  - **Mud**: Acts as a solid under low stress, but flows when sufficient force is applied.
  - **Toothpaste**: Remains in the tube without flowing until squeezed, then behaves like a fluid.
  - **ER fluids**: In the presence of an electric field, they can exhibit Bingham-like behavior, where they resist flow until the yield stress is overcome.
- **Applications:**
  The Bingham body model is widely applied in industries and fields where materials exhibit both solid and fluid characteristics. It is particularly useful for modeling the behavior of materials such as pastes, slurries, and smart fluids like ER and magnetorheological (MR) fluids. In civil engineering, the model is used to understand and predict the flow of concrete and mud. In food processing, it helps in the design and control of processes involving products like ketchup and toothpaste, which need to remain stable under low stress but flow easily when squeezed or pumped. The model is also crucial in the development of smart damping systems, where ER and MR fluids are used to create controllable resistance in devices like shock absorbers and vibration dampers.

Practical example: concrete flow in construction

- **Scenario**: During the pouring of concrete in a construction project, it is essential to ensure that the concrete flows smoothly into the forms but remains stable once set.
- **Solution**: The Bingham body model is used to predict the flow behavior of the concrete mixture.
- **Function**: Engineers calculate the yield stress of the concrete to ensure it will not flow until a sufficient force is applied, allowing it to be shaped as needed during pouring. Once the force is removed, the concrete stops flowing, maintaining its shape while it sets and hardens.
- **Outcome**: The model helps in optimizing the mix design and pouring process, ensuring that the concrete fills the forms correctly without excessive movement, leading to a stable and well-formed structure.

3.4 Newtonian viscosity and non-Newtonian viscosity

3.4.1 Newtonian viscosity

Newtonian viscosity refers to the behavior of fluids whose viscosity remains constant regardless of the applied shear rate or stress. In other words, these fluids have a linear relationship between shear stress and shear rate. The viscosity of a Newtonian fluid does not change with the rate at which it is stirred or deformed. Common examples of Newtonian fluids include water, air, and most simple liquids such as alcohol and gasoline.
- **Behavior**:
    - **Linear relationship**: In Newtonian fluids, the relationship between shear stress (the force applied) and shear rate (the resulting flow or deformation) is linear. This means that if you double the force applied, the flow rate doubles as well.
    - **Constant viscosity**: The viscosity, which is the measure of a fluid's resistance to flow, remains constant regardless of how fast or slow the fluid is stirred or sheared.
- **Examples**:
    - **Water**: This is a classic example of a Newtonian fluid. Whether you stir it slowly or quickly, its resistance to flow (viscosity) remains unchanged.
    - **Air**: As a gaseous Newtonian fluid, its viscosity remains consistent under varying rates of deformation.
    - **Simple oils**: Many oils, such as mineral oil, exhibit Newtonian behavior, maintaining consistent viscosity under different shear rates.
- **Applications**:
    - **Hydraulics**: In hydraulic systems, the use of Newtonian fluids such as mineral oils ensures predictable and consistent performance.
    - **Engineering calculations**: Many engineering and fluid dynamics calculations assume Newtonian behavior due to its simplicity and predictability.

3.4.2 Non-Newtonian viscosity

Non-Newtonian viscosity describes fluids whose viscosity changes when subjected to different shear rates or stresses. These fluids do not have a constant viscosity; instead, their flow behavior can be more complex, exhibiting either shear-thinning (where the viscosity decreases with increased shear rate) or shear-thickening (where the viscosity increases with increased shear rate). Other behaviors include viscoelasticity, where the fluid exhibits both viscous and elastic characteristics, and yield stress, where the fluid behaves like a solid until a certain stress threshold is reached [20].

Examples of non-Newtonian fluids include ketchup (shear-thinning), corn starch in water (shear-thickening), and toothpaste (yield stress).

–   **Behavior**:
    –   **Nonlinear relationship**: Non-Newtonian fluids exhibit a nonlinear relationship between shear stress and shear rate. Their viscosity changes based on the rate of applied stress or shear. These fluids can behave very differently under varying conditions, making them more complex to model and predict.
    –   **Variable viscosity**: The viscosity of non-Newtonian fluids can increase, decrease, or exhibit a combination of behaviors depending on the applied stress or shear rate.
–   **Applications**:
    –   **Cosmetics**: Non-Newtonian behavior in creams and gels allows them to be applied smoothly and stay in place once applied.
    –   **Food industry**: The control of flow properties in sauces, pastes, and dressings is crucial for both processing and consumer experience.
    –   **Pharmaceuticals**: The delivery and consistency of ointments and gels are managed by exploiting non-Newtonian properties.
    –   **Industrial processes**: Non-Newtonian fluids are used in various manufacturing processes, where control over material flow is critical.

3.5 Principal characteristics of electrorheological fluids

ER fluids are fascinating smart materials that exhibit a unique ability to rapidly change their mechanical properties in response to an applied electric field [21, 22]. This change is primarily characterized by a dramatic increase in viscosity, allowing the fluid to transition from a liquid state to a more solid-like state almost instantaneously. Let us explore the principal characteristics of ER fluids:

–   **Field-dependent viscosity:**
ER fluids exhibit field-dependent viscosity, a property where their viscosity changes in response to an applied electric field. Mechanistically, these fluids are composed of fine dielectric particles suspended in a nonconductive, insulating liquid. When an electric field is applied across the fluid, the dielectric particles become polarized, aligning themselves into chain-like structures or fibrils along the direction of the electric field. This alignment dramatically increases the fluid's resistance to flow, causing the viscosity to rise by several orders of magnitude, effectively transforming the fluid from a liquid into a semisolid or gel-like state. This controllable change in viscosity is particularly valuable in applications requiring precise regulation of fluid flow. For example, in adjustable dampers or clutches, ER fluids can be tuned to provide varying degrees of resistance, allowing for real-time adjustments in mechanical systems.

– **Reversibility:**

The viscosity change in ER fluids is fully reversible, a key characteristic that enhances their versatility in various applications. Mechanistically, when the electric field is removed, the dielectric particles within the fluid lose their polarization, leading to the disassembly of the chain structures that had formed under the field's influence. As a result, the fluid quickly returns to its original, low-viscosity state. This ability to switch back and forth between high and low viscosities is crucial in applications where dynamic control of fluid properties is required. For instance, in adaptive systems like active vibration control in vehicles, ER fluids can be continuously and rapidly adjusted to optimize ride comfort and handling, making them ideal for scenarios that demand quick and repeated changes in fluid behavior.

– **Yield stress:**

Yield stress is the minimum stress required to initiate flow in a fluid, marking the point where the fluid begins to deform and move. In ER fluids, the application of an electric field introduces a yield stress, below which the fluid behaves like a solid and resists any deformation. Only when the applied stress exceeds this threshold does the fluid transition into a flowing state. This characteristic is particularly important in applications that require controlled motion. For instance, in a braking system utilizing ER fluids, the fluid can maintain a solid-like state, preventing movement until a specific force is applied. This allows for a highly controllable braking mechanism, where the brake can be engaged or disengaged with precise control based on the stress applied, enhancing the system's responsiveness and reliability.

– **Response time:**

One of the most significant advantages of ER fluids is their rapid response time. The transition from a fluid to a semisolid state occurs within milliseconds of applying or removing an electric field. This quick response is driven by the fast polarization of the suspended particles and the rapid formation of chain-like structures within the fluid. The speed at which ER fluids can change their viscosity makes them ideal for real-time applications, such as active suspension systems in vehicles. In these systems, the fluid's damping characteristics can be adjusted on the fly, allowing the suspension to respond almost instantaneously to changes in road conditions or driving dynamics, thereby optimizing ride comfort and vehicle stability.

– **Dielectric properties:**

The dielectric properties of ER fluids are central to their function. The particles suspended in these fluids are typically made from materials such as polymers, ceramics, or metal oxides, all of which possess dielectric properties. When an electric field is applied, these particles become polarized, meaning they develop positive and negative charges on opposite sides. This polarization causes the particles to align and form chain-like structures along the direction of the electric field, which significantly increases the fluid's viscosity. The strength of this effect is directly related to the dielectric properties of the particles; the higher the dielectric constant, the more pronounced the change in viscosity. This relationship is crucial for enhancing the performance of ER fluids, as stronger dielectric properties lead to more effective and controllable changes in the fluid's behavior under an electric field.

– **Suspended particles:**
The suspended particles in ER fluids are typically micron-sized, ranging from 1 to 100 μm, and are dispersed in a nonconductive carrier liquid such as silicone oil, mineral oil, or other insulating fluids. The composition, concentration, and size of these particles are critical factors that determine how the fluid will behave under an electric field. The choice of particles, along with the carrier fluid, directly influences the ER fluid's performance characteristics, including its yield stress, the extent of viscosity change, and temperature stability. By carefully selecting and adjusting the concentration of these particles, engineers can tailor the ER fluid's properties to meet specific application needs, ensuring optimal performance in various environments and conditions.

– **Temperature sensitivity:**
ER fluids are sensitive to temperature variations, which can significantly impact their performance. At higher temperatures, the viscosity of the fluid may decrease and the ER effect can weaken, as increased thermal motion disrupts the formation of the particle chains that are essential for viscosity changes. Conversely, at very low temperatures, the carrier fluid can become more viscous or even freeze, reducing the fluid's ability to respond to an electric field. This temperature sensitivity means that in practical applications, it is crucial to carefully consider the operating temperature range of ER fluids. For example, in automotive applications, ER fluids must be capable of functioning effectively across a broad spectrum of temperatures, from the cold starts experienced in winter to the high operating temperatures encountered in summer. This ensures reliable performance and adaptability in diverse environmental conditions.

Applications in technology and engineering:

– **Adaptive suspension systems**: ER fluids are used in vehicle suspension systems to provide variable damping. By adjusting the electric field, the suspension system can instantly change its stiffness, offering a smoother ride on rough roads or increased stability during high-speed maneuvers [22].
– **Clutches and brakes**: ER fluids allow for the precise control of torque transfer in clutches and brakes. When an electric field is applied, the fluid's increased viscosity can lock the clutch or brake, transmitting torque or stopping motion as needed.

- **Vibration damping**: In structures or machinery, ER fluids are used in dampers to reduce vibrations. The damping force can be adjusted in real time to counteract vibrations from engines, earthquakes, or other dynamic forces.
- **Haptic devices**: ER fluids are utilized in haptic feedback devices, where the fluid's resistance can be controlled to simulate different textures or forces, enhancing the user experience in virtual reality or gaming systems.

3.6 The electrorheological phenomenon

The ER phenomenon refers to the dramatic change in the viscosity of certain fluids when exposed to an electric field. In ER fluids, dielectric particles suspended in a nonconductive liquid become polarized when an electric field is applied. This polarization causes the particles to align and form chain-like structures or fibrils along the direction of the electric field. As these structures form, the fluid's resistance to flow increases significantly, effectively turning it from a liquid into a semisolid or gel-like state. When the electric field is removed, the particles lose their polarization, the chains disassemble, and the fluid returns to its original low-viscosity state [22]. This reversible change in viscosity allows for dynamic control of the fluid's flow properties, making ER fluids useful in applications such as adjustable dampers, clutches, and vibration control systems.

3.7 Charge migration mechanism for the dispersed phase

The charge migration mechanism for the dispersed phase in ER fluids involves the movement of electrical charges within the suspended particles when an electric field is applied. In ER fluids, the dispersed phase consists of dielectric particles suspended in a nonconductive liquid. When an electric field is introduced, these particles become polarized – meaning positive and negative charges migrate to opposite sides of the particles. This polarization causes the particles to align and form chain-like structures along the direction of the electric field [22]. The formation of these structures increases the fluid's viscosity by disrupting its flow. When the electric field is removed, the charges dissipate, and the particles return to their random orientation, causing the fluid to revert to its original, lower viscosity state.

3.8 Electrorheological fluid domain

The ER fluid domain refers to the field of study and application involving fluids whose rheological (flow) properties can be controlled by applying an electric field. This domain encompasses the understanding of how ER fluids behave under electric fields, including their ability to rapidly change viscosity from a liquid to a semisolid state. It involves exploring the underlying mechanisms, such as particle polarization and alignment, which drive these changes. The domain also covers the design and optimization of ER fluids for various applications, including adaptive suspension systems, clutches, and brakes, where precise control over fluid properties is crucial. Researchers and engineers in this field work on improving the performance, responsiveness, and stability of ER fluids to enhance their effectiveness in practical applications.

The ER fluid domain encompasses various applications where the fluid's viscosity can be dynamically controlled by an electric field. Key applications include:

- **Adaptive suspension systems**: ER fluids are used in vehicle suspensions to adjust damping in real time based on road conditions, improving ride comfort and vehicle stability.
- **Clutches and brakes**: ER fluids provide precise control in clutches and braking systems, allowing for smooth engagement and disengagement by altering fluid viscosity with an electric field.
- **Vibration damping**: ER fluids are employed in dampers to control vibrations in machinery and structures, enhancing performance and durability by adjusting the fluid's resistance to vibrations.
- **Haptic feedback devices**: In interactive devices, ER fluids offer variable resistance to simulate different textures and forces, enhancing the tactile feedback in virtual reality and gaming systems.
- **Active vibration control**: ER fluids help manage vibrations in structures or vehicles by adjusting fluid viscosity and improving comfort and structural integrity through effective vibration damping.

3.9 Electrorheological fluid actuators

ER fluid actuators are devices that use ER fluids to control mechanical movements or forces by varying the fluid's viscosity through an electric field. These actuators operate by applying an electric field to ER fluids, which causes the suspended dielectric particles to align and form chain-like structures. This alignment significantly increases the fluid's viscosity, effectively turning it from a liquid into a semisolid state. As a result, the actuator can precisely control resistance, force, or movement based on the electric field's strength.

ER fluid actuators can be classified into several types based on their design and application. Here are the main types:

3.9.1 ER fluid dampers

ER fluid dampers are devices that use ER fluids to provide adjustable damping by varying the fluid's viscosity through an electric field.

Working principle

ER fluid dampers operate by applying an electric field to the ER fluid contained within the damper. This field causes the dielectric particles in the fluid to polarize and align into chain-like structures. The formation of these structures significantly increases the fluid's viscosity, which in turn alters the damping characteristics of the damper [23]. When the electric field is removed, the particles lose their alignment, and the fluid's viscosity returns to its original, lower state. This ability to adjust the viscosity in real time allows the damper to modify its damping force dynamically, responding to changing conditions or control inputs.

Description of components

– **ER fluid chamber:**
   – **Description:** The chamber contains the ER fluid and is typically designed to allow for fluid movement or flow between different sections of the damper.
   – **Function:** It holds the ER fluid and ensures that it is exposed to the electric field when required.
– **Electrodes:**
   – **Description:** Electrodes are placed within or around the ER fluid chamber and are used to apply the electric field to the fluid.
   – **Function:** They generate the electric field that causes the dielectric particles in the ER fluid to align, thereby changing the fluid's viscosity.
– **Piston and cylinder:**
   – **Description:** The piston moves within the cylinder, and the motion is resisted by the ER fluid's viscosity.
   – **Function:** They are part of the mechanical system that translates the changing viscosity of the ER fluid into damping forces [24]. The piston's movement through the ER fluid in the cylinder determines the damping effect.
– **Control system:**
   – **Description:** This system regulates the electric field applied to the ER fluid, based on inputs or conditions.
   – **Function:** It adjusts the strength of the electric field to control the viscosity of the ER fluid, thus tuning the damper's performance in real time.
– **Application**: This is commonly used in vehicle suspension systems and vibration control in machinery.

3.9.2 ER fluid clutches

ER fluid clutches are devices that use ER fluids to control torque transmission by varying the fluid's viscosity with an electric field [25]. Here is a brief explanation of their working principle and key components:

Working principle
ER fluid clutches operate by applying an electric field to the ER fluid contained within the clutch assembly. The application of the electric field causes the dielectric particles in the ER fluid to align and form chain-like structures. This alignment increases the fluid's viscosity, which enhances the clutch's ability to transmit torque. When the electric field is removed, the particles lose their alignment, and the fluid's viscosity decreases, allowing for disengagement or reduced torque transmission. This variable viscosity allows for smooth and precise control of torque.

Description of components
- **ER fluid chamber:**
  - **Description:** The chamber holds the ER fluid and is typically designed to allow for interaction between the fluid and the clutch's other components.
  - **Function:** It contains the ER fluid that changes viscosity in response to the electric field.
- **Electrodes:**
  - **Description:** Electrodes are positioned within or around the ER fluid chamber to apply the electric field.
  - **Function:** They create the electric field that causes the dielectric particles in the ER fluid to align, thereby modifying the fluid's viscosity and affecting the clutch's engagement.
- **Clutch plates:**
  - **Description:** These are typically a pair of friction surfaces between which the ER fluid is sandwiched.
  - **Function:** The clutch plates interact with the ER fluid to transmit or disengage torque based on the fluid's viscosity.
- **Control system:**
  - **Description:** The system that regulates the electric field is applied to the ER fluid.
  - **Function:** It adjusts the electric field strength to control the viscosity of the ER fluid, thereby regulating the amount of torque transmitted through the clutch.
- **Input and output shafts:**
  - **Description:** They are shafts that transfer torque to and from the clutch.

- **Function:** The input shaft provides the driving force, and the output shaft receives the transmitted torque, with the clutch controlling the engagement between them.
- **Application:** This is found in automotive transmissions and industrial machinery.

3.9.3 ER fluid brakes

ER fluid brakes use ER fluids to control braking force by varying the fluid's viscosity with an electric field. Here is a brief overview of their working principle and key components.

Working principle
ER fluid brakes operate by applying an electric field to the ER fluid contained within the brake assembly. The electric field causes dielectric particles in the ER fluid to align and form chain-like structures, significantly increasing the fluid's viscosity. This increased viscosity enhances the braking force by creating greater resistance to movement. When the electric field is removed, the particle chains disassemble, and the fluid's viscosity decreases, reducing the braking force or allowing the brake to disengage. This allows for precise and adjustable control of braking force.

Description of components
- **ER fluid chamber:**
  - **Description:** This chamber houses the ER fluid and is part of the brake system where the fluid's viscosity is manipulated.
  - **Function:** It contains the ER fluid, which changes its viscosity in response to the electric field, affecting the braking performance.
- **Electrodes:**
  - **Description:** Electrodes are positioned within or around the ER fluid chamber.
  - **Function:** They apply an electric field to the ER fluid, causing the dielectric particles to align and modify the fluid's viscosity.
- **Brake Disks or drums:**
  - **Description:** These are the friction surfaces that interact with the ER fluid.
  - **Function:** They work with the ER fluid to provide the braking force. The increased viscosity of the fluid creates greater frictional resistance against the disks or drums.
- **Control system:**
  - **Description:** This is the system that regulates the electric field applied to the ER fluid.

- **Function:** It adjusts the strength of the electric field to control the viscosity of the ER fluid, thereby controlling the braking force applied.
- **Actuator:**
  - **Description:** The component that engages or disengages the brake based on the viscosity of the ER fluid.
  - **Function:** It moves the brake components into or out of contact with the ER fluid, adjusting the braking force.
- **Application:** This is used in braking systems for vehicles and industrial equipment.

3.9.4 ER fluid valves

ER fluid valves use ER fluids to regulate the fluid flow by varying the fluid's viscosity in response to an electric field. Here is a brief explanation of their working principle and key components:

Working principle

ER fluid valves operate by applying an electric field to the ER fluid contained within the valve. When the electric field is applied, the dielectric particles in the ER fluid align and form chain-like structures, which significantly increase the fluid's viscosity. This increased viscosity changes the flow characteristics of the fluid through the valve. When the electric field is removed, the viscosity of the ER fluid decreases, allowing for normal flow through the valve. This dynamic control enables precise regulation of fluid flow rates and pressures.

Description of components
- **ER fluid chamber:**
  - **Description:** This is the chamber where the ER fluid is contained and where flow regulation occurs.
  - **Function:** It holds the ER fluid and allows the application of an electric field to alter the fluid's viscosity.
- **Electrodes:**
  - **Description:** Electrodes are positioned within or around the ER fluid chamber.
  - **Function:** They generate the electric field that causes the dielectric particles in the ER fluid to align, thus changing the fluid's viscosity and affecting flow control.
- **Flow passage:**
  - **Description:** This is the channel or opening through which the fluid flows.
  - **Function:** The flow passage is affected by the viscosity of the ER fluid, which is controlled by the electric field. Adjusting the viscosity alters the flow rate or pressure of the fluid passing through the valve.

- **Control system:**
  - **Description:** This is the system that regulates the electric field applied to the ER fluid.
  - **Function:** It adjusts the strength of the electric field to control the viscosity of the ER fluid, thereby managing the flow characteristics of the valve.
- **Actuator:**
  - **Description:** The mechanism that operates the valve based on the viscosity of the ER fluid.
  - **Function:** It adjusts the position of the valve components to control the flow passage and regulate the fluid flow.
- **Application:** This is employed in hydraulic and pneumatic systems for precise control of fluid flow.

3.9.5 ER fluid actuators for precision control

ER fluid actuators for precision control are devices that use ER fluids to achieve fine and accurate control of mechanical movements. Here is a brief overview of their working principle and key components:

Working principle
ER fluid actuators work by applying an electric field to the ER fluid contained within the actuator. The electric field causes the dielectric particles in the ER fluid to align and form chain-like structures, which increases the fluid's viscosity. This change in viscosity translates into varying levels of resistance or force within the actuator. By adjusting the electric field, the actuator can precisely control the force or position of a moving component. When the electric field is removed, the viscosity of the ER fluid decreases, allowing the actuator to return to its original state.

Description of components
- **ER fluid chamber:**
  - **Description:** This is the chamber where the ER fluid is housed.
  - **Function:** This contains the ER fluid and is designed to allow the electric field to interact with the fluid.
- **Electrodes:**
  - **Description:** They are positioned within or around the ER fluid chamber.
  - **Function:** Apply the electric field to the ER fluid, causing changes in viscosity that affect the actuator's performance.
- **Actuating element (piston or rotor):**
  - **Description:** This is the component that moves in response to changes in the ER fluid's viscosity.

- **Function:** This converts the variable viscosity of the ER fluid into mechanical motion or force. The movement of the piston or rotor is controlled by the viscosity changes in the fluid.
- **Control system:**
  - **Description:** This manages the electric field applied to the ER fluid.
  - **Function:** This adjusts the strength of the electric field to control the viscosity of the ER fluid, thereby precisely regulating the actuator's movement or force.
- **Feedback mechanism:**
  - **Description:** Sensors or devices that monitor the actuator's position or force.
  - **Function:** This provides real-time feedback to the control system, allowing for accurate adjustments and fine control of the actuator's performance.
- **Application:** This is used in robotic systems, aerospace applications, and advanced manufacturing processes.

3.10 Electrorheological fluid design parameter

ER fluid design parameters are key factors that determine the performance and effectiveness of ER fluids in various applications. Here is a brief overview of the essential design parameters:

- **Particle size and concentration:**
  - **Description:** The size and concentration of dielectric particles suspended in the ER fluid.
  - **Impact:** Larger particles or higher concentrations generally enhance the fluid's ability to form chain-like structures and increase viscosity under an electric field. However, very large particles may lead to sedimentation or clogging.
- **Dielectric properties:**
  - **Description:** The ability of the particles and carrier fluid to become polarized in an electric field.
  - **Impact:** Higher dielectric constants of the particles improve the fluid's responsiveness to the electric field, resulting in more pronounced changes in viscosity.
- **Carrier Fluid:**
  - **Description:** The nonconductive liquid in which the dielectric particles are suspended (e.g., silicone oil and mineral oil).
  - **Impact:** The choice of carrier fluid affects the fluid's overall viscosity, temperature stability, and compatibility with the suspended particles. The carrier fluid should maintain its properties over the operational temperature range.
- **Electric field strength:**
  - **Description:** The intensity of the electric field applied to the ER fluid.

- **Impact:** The strength of the electric field directly influences the degree of viscosity change in the fluid. Higher field strengths typically result in greater viscosity increases.
- **Temperature stability:**
    - **Description:** The fluid's ability to maintain performance across different temperatures.
    - **Impact:** Temperature fluctuations can affect the viscosity of the ER fluid and the carrier fluid's properties. Design parameters should ensure that the fluid remains effective in the intended operational temperature range.
- **Response time:**
    - **Description:** The time it takes for the ER fluid to change its viscosity in response to the electric field.
    - **Impact:** Faster response times are crucial for applications requiring rapid adjustments in fluid properties. The design should minimize lag between field application and viscosity change.
- **Shear strength:**
    - **Description:** The fluid's ability to resist shear forces when subjected to stress.
    - **Impact:** High shear strength helps in maintaining the structural integrity of the fluid and ensures consistent performance under dynamic conditions.

3.11 Applications of electrorheological fluids

Applications of ER fluids leverage their ability to rapidly change viscosity in response to an electric field. Here is a brief overview of some key applications:
- **Clutches:**
    - **Description:** ER fluids are used in clutches to modulate torque transmission.
    - **Application:** By adjusting the electric field, the viscosity of the ER fluid changes, allowing the clutch to engage or disengage smoothly and precisely. This is useful in automotive and industrial machinery where variable torque control is needed.
- **Dampers:**
    - **Description:** ER fluids are employed in shock absorbers and suspension systems.
    - **Application:** In response to an electric field, the viscosity of the fluid changes, altering the damping characteristics. This provides adjustable ride comfort and handling in vehicles and helps control vibrations in machinery.
- **Brakes:**
    - **Description:** ER fluids are used in braking systems to control the braking force.

  – **Application:** The fluid's viscosity increases with the electric field, enhancing the braking force. This is used in applications where variable braking force is needed, such as in automotive systems and robotics.
- **Valves:**
  – **Description:** ER fluids are used in control valves to regulate fluid flow.
  – **Application:** The viscosity of the fluid changes with the electric field, which alters the flow rate through the valve. This is useful in hydraulic systems, fluid control, and industrial processes where precise flow regulation is required.
- **Actuators:**
  – **Description:** ER fluids are utilized in actuators for precision control.
  – **Application:** By varying the electric field, the actuator can achieve precise control over force and movement. This is valuable in robotics, aerospace systems, and various automation technologies where fine control is essential.
- **Aerospace and automotive systems:**
  – **Description:** ER fluids are used in advanced aerospace and automotive systems.
  – **Application:** In aerospace, ER fluids help manage vibration and control systems. In automotive applications, they enhance the ride quality and handling by providing adaptive suspension systems.

3.12 Summary

This chapter explores the fundamental properties and applications of these smart materials that change viscosity in response to an electric field. It covers the composition of suspensions and ER fluids, the Bingham body model for describing their flow behavior, and contrasts Newtonian and non-Newtonian viscosity. Key characteristics of ER fluids, including their field-dependent viscosity, reversibility, and rapid response, are detailed along with the ER phenomenon and charge migration mechanism. The ER fluid domain highlights the various conditions under which ER fluids operate, while ER fluid actuators and design parameters emphasize their precise control and customization. Applications span across clutches, dampers, brakes, and other systems where ER fluids provide adaptable and efficient performance.

Review questions

1. What are ER fluids and how do they differ from other smart materials?
2. Describe the basic composition of an ER fluid.
3. How does the viscosity of ER fluids change in response to an electric field?
4. What role do suspended particles play in ER fluids?

5. Why is the choice of carrier fluid important in ER fluids?
6. How does particle size affect the behavior of ER fluids?
7. What is the Bingham body model and how does it describe the flow of ER fluids?
8. How does the Bingham body model differentiate between solid-like and fluid-like behavior?
9. What are the key parameters in the Bingham body model for ER fluids?
10. Define Newtonian viscosity and provide an example of a Newtonian fluid.
11. What distinguishes non-Newtonian viscosity from Newtonian viscosity?
12. How do ER fluids exhibit non-Newtonian behavior?
13. What are the principal characteristics of ER fluids?
14. How does the electric field strength influence ER fluid properties?
15. What is meant by the reversibility of ER fluids?
16. Describe the ER phenomenon and its effects on fluid viscosity.
17. How do dielectric particles in ER fluids contribute to viscosity changes?
18. What factors affect the effectiveness of the ER phenomenon?
19. Explain the charge migration mechanism in ER fluids.
20. How does the alignment of dielectric particles impact the fluid's viscosity?
21. What is the role of polarization in the charge migration mechanism?
22. What is the ER fluid domain and why is it important?
23. How do environmental factors influence the performance of ER fluids?
24. What are the operational limitations of ER fluids in different domains?
25. What are ER fluid actuators and how do they function?
26. Describe the working principle of ER fluid actuators.
27. How does the electric field control the movement of an ER fluid actuator?
28. What are the key design parameters for ER fluids?
29. How does particle concentration affect the performance of ER fluids?
30. What considerations are necessary for selecting the carrier fluid in ER fluids?
31. List some common applications of ER fluids.
32. How are ER fluids used in automotive clutches?
33. Describe the use of ER fluids in damping systems.
34. What role do ER fluids play in braking systems?
35. How do ER fluids function in control valves?
36. Explain how ER fluids are utilized in precision actuators.
37. What are the benefits of using ER fluids in aerospace systems?
38. How do ER fluids enhance the ride quality in vehicles?
39. What are the challenges of using ER fluids in extreme temperature conditions?
40. Discuss the advantages of ER fluids in industrial machinery.
41. What is the significance of the field-dependent viscosity in ER fluids?
42. How does temperature affect the viscosity and performance of ER fluids?
43. What are the typical materials used for dielectric particles in ER fluids?
44. Describe the typical structure of an ER fluid system in a practical application.

45. How does the response time of ER fluids impact their application in real-time systems?
46. What are the main differences between ER fluids and MR fluids?
47. How can the performance of ER fluids be optimized for specific applications?
48. What are the potential future developments for ER fluids?
49. How do ER fluid design parameters influence their practical applications?
50. Discuss the impact of electric field strength on the efficiency of ER fluid devices.

4 Piezoelectric smart materials

4.1 Introduction

Piezoelectric smart materials are a class of materials that generate an electrical charge in response to mechanical stress and vice versa. This unique property, known as the piezoelectric effect, allows these materials to convert mechanical energy into electrical energy and electrical energy into mechanical energy. Piezoelectric materials are typically crystals or ceramics that exhibit this behavior due to their internal electric dipole structure, which aligns under applied stress, creating an electric charge. These materials are widely used in various applications, including sensors, actuators, and energy-harvesting devices. In sensors, piezoelectric materials convert physical changes, such as pressure or vibrations, into electrical signals that can be measured and analyzed. As actuators, they can convert electrical signals into precise mechanical movements, making them essential in high-precision and responsiveness applications [26]. The ability to harvest energy from mechanical vibrations or pressure changes also positions piezoelectric materials as crucial components in sustainable energy solutions. The versatility and functionality of piezoelectric smart materials make them integral to modern technological advancements across industries such as aerospace, automotive, electronics, and medical devices.

4.2 Background

Piezoelectric smart materials have a rich history rooted in the discovery of the piezoelectric effect, first observed by the French physicists Pierre and Jacques Curie in 1880. They found that certain crystals, such as quartz, produce an electric charge when subjected to mechanical stress. This phenomenon arises from the alignment of internal electric dipoles within the crystal lattice under applied stress, resulting in an electrical charge [27].

The early twentieth century saw the development and application of piezoelectric materials in various fields, particularly in the electronics and telecommunications industries. The discovery of materials such as barium titanate and lead zirconate titanate (PZT) further expanded the use of piezoelectric materials due to their improved piezoelectric properties. These advancements enabled the creation of more sensitive sensors and more effective actuators.

In the latter half of the twentieth century, piezoelectric materials gained prominence in precision engineering, aerospace, and medical applications. Their ability to convert mechanical energy into electrical signals and vice versa found practical use in devices such as ultrasonic transducers, accelerometers, and vibration sensors. In-

https://doi.org/10.1515/9783111379623-004

novations in material science, including the development of piezoelectric polymers and composites, broadened the scope of piezoelectric applications.

Today, piezoelectric smart materials are integral to numerous technologies, from everyday electronic devices to sophisticated industrial systems. They are used in sensors for detecting pressure, strain, and vibrations, as well as in actuators for precise control in robotics and aerospace systems. Their applications also extend to energy harvesting, where they convert mechanical vibrations into electrical power, contributing to advancements in sustainable and autonomous systems. The ongoing research and development in piezoelectric materials continue to drive innovations, expanding their capabilities and applications across various industries.

4.3 Electrostriction

Electrostriction is a phenomenon observed in piezoelectric smart materials where the material changes the shape or size in response to an applied electric field. Unlike piezoelectricity, which involves the alignment of internal dipoles leading to reversible mechanical deformation, electrostriction results in a uniform, nonlinear deformation of the material [28]. This deformation is due to the interaction of the electric field with the material's dielectric properties, causing a change in polarization and, consequently, a change in shape.

Key points
– **Mechanism:** In piezoelectric materials, electrostriction manifests as a result of the material's response to an electric field, leading to uniform deformation. This effect is quadratic with respect to the field strength, meaning that the amount of deformation increases with the square of the applied electric field.
– **Characteristics:** Electrostrictive behavior in piezoelectric materials is characterized by stable and predictable shape changes, which are advantageous for applications requiring precise and consistent control over mechanical movement.
– **Applications:** Electrostriction enhances the functionality of piezoelectric smart materials in various applications, such as in actuators and sensors where large, precise deformations are beneficial. This capability is instrumental in advanced technologies requiring fine control over mechanical responses, such as adaptive optics, precision engineering, and vibration control systems.

4.4 Pyroelectricity

Pyroelectricity is the phenomenon where certain materials generate an electric charge in response to a change in temperature. Unlike piezoelectricity, which is caused by mechanical stress, pyroelectricity arises from temperature-induced changes in the material's polarization [29]. As the temperature changes, the alignment of electric dipoles within the material shifts, leading to a variation in the electric field and resulting in a measurable voltage.

Key points
- **Mechanism:** Pyroelectric materials generate an electric charge due to temperature fluctuations that alter their polarization. This effect is reversible, with the charge varying directly with temperature changes.
- **Characteristics:** Pyroelectric materials exhibit a temperature-dependent electric polarization, which creates an electric field when the temperature changes. This effect is particularly useful for detecting thermal variations.
- **Applications:** Pyroelectric materials are used in infrared sensors, thermal imaging devices, and temperature measurement systems, where their ability to detect changes in temperature with high sensitivity is beneficial.

4.5 Piezoelectricity

Piezoelectricity is the ability of certain materials to generate an electric charge in response to applied mechanical stress. This effect occurs due to the alignment of internal electric dipoles within the material's crystal lattice when it is deformed. The application of stress causes a shift in the dipoles, leading to the generation of an electric voltage across the material [13].

Key points
- **Mechanism:** When mechanical stress is applied to a piezoelectric material, the internal dipoles align, producing an electric charge. Conversely, applying an electric field can cause the material to change its shape.
- **Characteristics:** Piezoelectricity is reversible, meaning that mechanical stress creates an electric charge, and an electric field can induce mechanical deformation. This behavior is highly dependent on the material's crystal structure.
- **Applications:** Piezoelectric materials are used in a variety of applications, including sensors (e.g., pressure sensors), actuators (e.g., precise movement devices), and energy harvesters (e.g., converting vibrations into electrical energy).

4.6 Industrial piezoelectric materials

Industrial piezoelectric materials are specialized materials used in various industrial applications for their ability to convert mechanical stress into electrical energy and vice versa. They come in several types, each suited for different applications based on their properties.

Types

1. **Ceramics:**
   - **Examples:** PZT and barium titanate.
   - **Characteristics:** High piezoelectric coefficients and strong electrical response. They are commonly used in sensors and actuators due to their robust performance and high sensitivity.
2. **Crystals:**
   - **Examples:** Quartz and lithium niobate.
   - **Characteristics:** Naturally occurring or synthetically grown, these materials have stable piezoelectric properties and are used in applications requiring precise frequency control, such as in oscillators and frequency filters.
3. **Polymers:**
   - **Examples:** Polyvinylidene fluoride (PVDF) and polyamide.
   - **Characteristics:** Flexible and lightweight with good piezoelectric response. They are used in applications where flexibility and conformability are required, such as in flexible sensors and actuators.
4. **Composites:**
   - **Examples:** Piezoelectric ceramic–polymer composites.
   - **Characteristics:** Combine the properties of ceramics and polymers to enhance performance. They offer improved mechanical flexibility and durability, making them suitable for complex and demanding applications.

Applications

- **Sensors:** Used in pressure sensors, accelerometers, and vibration sensors to detect mechanical changes and convert them into electrical signals.
- **Actuators:** Employed in precision movement devices, such as in medical ultrasound equipment and automotive components, where precise control and actuation are needed.
- **Energy harvesting:** Integrated into systems to capture and convert mechanical energy (e.g., vibrations) into electrical energy for powering small devices or sensors.
- **Frequency control:** Utilized in quartz crystals and oscillators for stable and precise frequency control in electronic devices.

4.7 PZT film

PZT film is a thin layer of a piezoelectric ceramic material widely used for its high piezoelectric and dielectric properties. PZT is a perovskite material that can convert mechanical energy into electrical energy and vice versa, making it highly suitable for various industrial and technological applications, especially in miniaturized devices.

Detailed characteristics
- **High piezoelectric coefficient:** PZT films exhibit strong piezoelectric effects, generating substantial electrical charge in response to mechanical stress. This high sensitivity makes them ideal for applications requiring precise control and responsiveness.
- **Flexible deposition methods:** PZT films can be deposited on substrates using methods such as:
    - **Sol–gel processing:** A chemical technique that produces uniform, high-quality films.
    - **Sputtering:** A physical vapor deposition method that allows for controlled film thickness and composition.
    - **Chemical vapor deposition:** Enables large-scale production with good adhesion properties.
- **Tailorability:** The properties of PZT films can be adjusted by varying the composition (e.g., the ratio of lead, zirconium, and titanium), processing conditions, and doping with other elements. This allows for customization according to specific application requirements.

Applications
1. **Microelectromechanical systems (MEMS):**
    - PZT films are integral in MEMS devices, where they function as sensors and actuators due to their ability to generate mechanical displacement with high precision.
2. **Ultrasonic transducers:**
    - In medical imaging, PZT films are used in ultrasound transducers for their ability to produce high-frequency sound waves and detect reflected waves, enabling clear and detailed imaging.
3. **Precision positioning:**
    - In nanopositioning devices, such as those used in semiconductor manufacturing, PZT films provide the fine control necessary for moving components with nanometer-level precision.

4. **Energy harvesting:**
   - PZT films are used in small energy harvesters that capture ambient vibrations and convert them into electrical energy, powering low-energy devices such as wireless sensors.
5. **Optical devices:**
   - In adaptive optics, PZT films control the shape of mirrors or lenses by applying precise deformation to correct for optical aberrations in real time.
6. **Consumer electronics:**
   - PZT films are utilized in devices like autofocus systems in cameras, haptic feedback mechanisms in touchscreens, and high-precision inkjet printers.

Advantages of PZT films
- **Miniaturization:** Thin films allow for compact, lightweight designs, crucial in portable and space-constrained devices.
- **High efficiency:** PZT's strong electromechanical coupling ensures efficient energy conversion.
- **Durability and stability:** PZT films maintain their performance over long periods, even under harsh operating conditions.

A practical example of PZT films in action is their use in ultrasonic medical imaging (ultrasound transducers).

Application in ultrasound transducers
- **Working principle:** In medical ultrasound devices, PZT films serve as the core component of the transducers. When an alternating electric field is applied to the PZT film, it undergoes rapid expansion and contraction, generating high-frequency sound waves. These waves travel through the body and reflect off tissues and organs. The same PZT film then detects the reflected waves, which convert them back into electrical signals to create detailed images of internal body structures.
- **Benefits in medical imaging:**
  - **High resolution:** PZT films produce precise sound waves, leading to high-resolution images essential for accurate diagnostics.
  - **Real-time imaging:** The fast response of PZT allows for real-time imaging, which is critical in monitoring dynamic processes like blood flow.
  - **Compact and lightweight design:** The thin film nature of PZT makes the transducers compact, lightweight, and easy to handle, which is crucial for handheld ultrasound probes.

4.8 PVDF film

PVDF film is a prominent piezoelectric polymer with unique properties that make it suitable for a wide range of applications. Unlike traditional piezoelectric ceramics such as PZT, PVDF offers the advantage of flexibility, durability, and ease of processing, making it a popular choice in applications where flexibility and large-area coverage are required.

Characteristics
1. **High flexibility:**
   - Unlike ceramic piezoelectric materials that are brittle and rigid, PVDF is a flexible polymer that can be bent, stretched, and twisted without breaking. This flexibility allows it to be integrated into wearable devices, curved surfaces, and dynamic environments.
2. **Piezoelectric and pyroelectric properties:**
   - PVDF exhibits both piezoelectric and pyroelectric properties. When subjected to mechanical stress, it generates an electric charge (piezoelectric effect), and when exposed to temperature changes, it produces a voltage (pyroelectric effect). This dual functionality expands its application potential in both sensing and energy harvesting.
3. **Lightweight and thin:**
   - PVDF films are extremely lightweight and can be made very thin, down to micrometer scales, which is essential for applications in compact, portable, or wearable devices.
4. **Chemical and environmental resistance:**
   - PVDF is highly resistant to chemicals, moisture, and UV radiation. It remains stable over a wide range of temperatures, from −40 °C to +150 °C, making it suitable for harsh environmental conditions [30].
5. **Good sensitivity:**
   - While the piezoelectric response of PVDF is lower compared to ceramics, its sensitivity is still adequate for many applications, especially where flexibility and large-area sensing are prioritized.

Applications
1. **Flexible and wearable electronics:**
   - PVDF films are increasingly used in wearable health monitoring devices, where they measure parameters like heart rate, respiratory rate, and movement. For instance, PVDF can be embedded into smart fabrics or patches to monitor vital signs in real time.

2. **Biomedical sensors:**
   - In the medical field, PVDF is used in pressure sensors and catheters for minimally invasive surgeries. Its flexibility allows it to conform to body contours, providing accurate pressure measurements without causing discomfort.
3. **Energy harvesting:**
   - PVDF is utilized in energy-harvesting applications, where it converts ambient mechanical vibrations (e.g., from human motion or machinery) into electrical energy. This energy can be used to power low-energy devices like wireless sensors, enabling self-sustaining systems.
4. **Acoustic and vibration sensors:**
   - PVDF films are used in microphones, hydrophones, and vibration sensors due to their ability to detect sound waves and vibrations over a wide frequency range. Their lightweight nature and flexibility make them suitable for underwater applications and large-area acoustic arrays.
5. **Structural health monitoring:**
   - PVDF films are integrated into structures like bridges and buildings to monitor strain, stress, and vibrations. They provide real-time data on structural health, allowing for preventive maintenance and early detection of potential failures.

A practical example of PVDF film usage is in wearable health monitors, particularly those designed for tracking heart rate, respiration, and body movements. These devices leverage the piezoelectric properties of PVDF, combined with its flexibility, to create sensors that are comfortable, accurate, and capable of continuous monitoring.

How it works
- **Heart rate monitoring:**
   - PVDF films are embedded into straps, patches, or even clothing that contacts the skin. As the heart beats, the resulting pulse generates minute mechanical pressure changes on the PVDF film. The film responds to this pressure by generating a corresponding electrical signal due to its piezoelectric nature. These signals are processed to determine the heart rate in real time.
- **Respiration tracking:**
   - During breathing, the chest and abdomen expand and contract. PVDF sensors, placed near these areas, detect the strain caused by this motion. As the body moves with each breath, the PVDF film stretches and compresses, converting mechanical deformation into electrical signals. The frequency and intensity of these signals are used to calculate respiration rates and monitor breathing patterns.

- **Motion and activity tracking:**
  - PVDF films are sensitive enough to detect even subtle body movements. Whether embedded in smart clothing, fitness bands, or adhesive patches, these films can measure steps, postural changes, and overall activity levels. The flexibility of PVDF allows the sensors to bend and stretch with the body, maintaining accuracy without restricting movement.

Benefits of using PVDF films in wearable monitors

1. **Flexibility and comfort:**
   - The lightweight and bendable nature of PVDF films makes them comfortable for prolonged use, ensuring that wearers do not experience discomfort, even during extended periods of activity or sleep.
2. **High sensitivity:**
   - Despite their thin profile, PVDF films can capture even subtle physiological changes, enabling precise monitoring of vital signs.
3. **Versatile integration:**
   - PVDF films can be easily integrated into various wearable formats, from wristbands and patches to smart textiles. This adaptability allows manufacturers to design health monitoring devices that are unobtrusive and easy to wear in daily life.
4. **Continuous and real-time monitoring:**
   - The electrical output of PVDF sensors can be continuously measured, making them ideal for real-time monitoring applications in healthcare, fitness, and wellness. This continuous data can be transmitted to smartphones or health platforms for further analysis, enabling early detection of anomalies.

Application in healthcare and fitness

- **Fitness tracking:**
  - In fitness devices, PVDF films measure activity levels, steps, and exercise intensity. The sensors can detect the heart rate changes and respiration patterns during workouts, providing users with insights into their cardiovascular performance and fitness levels.
- **Healthcare monitoring:**
  - PVDF-based wearable monitors are used in chronic disease management. For instance, they can monitor respiration in patients with sleep track heart rate variability in individuals with cardiovascular conditions. Continuous data collection allows for better patient management and personalized healthcare interventions.
- **Remote patient monitoring:**
  - PVDF sensors are also used in remote health monitoring systems where patients can be monitored outside of hospital settings. These wearable devices

collect vital data that can be sent to healthcare providers for analysis, enabling timely medical interventions.

Practical example: smart shirts for health monitoring
An example of a product using PVDF technology is a **smart shirt** designed for athletes or patients. The shirt has PVDF sensors woven into the fabric that can monitor the heart rate, respiration, and even posture. As the person wears the shirt, the embedded PVDF films sense body movements and physiological changes, transmitting the data wirelessly to a smartphone or computer for analysis. The flexibility of the material ensures that the sensors remain effective without being felt by the wearer, offering a seamless health monitoring experience.

In summary, PVDF films offer an ideal solution for wearable health monitoring due to their flexibility, high sensitivity, and ability to be integrated into everyday clothing and accessories. This makes them an essential component in the future of continuous health monitoring and personalized healthcare.

4.9 Properties of commercial piezoelectric materials

Commercial piezoelectric materials possess several key properties that make them useful in a wide range of applications. These materials can be categorized into ceramics, polymers, and composites, each having specific characteristics. Below are some of the essential properties of commercial piezoelectric materials:

1. **Piezoelectric coefficient:**
   - The piezoelectric coefficient quantifies the material's ability to convert mechanical stress into electrical charge and vice versa. Higher coefficients indicate stronger piezoelectric effects.
   - Ceramics like PZT have high piezoelectric coefficients, making them ideal for actuators and sensors.
   - Polymers like PVDF have lower coefficients but are more flexible, making them suitable for wearable applications.
2. **Dielectric constant:**
   - The dielectric constant measures the material's ability to store electrical energy under an electric field.
   - High dielectric constants in materials like PZT enhance their sensitivity and performance in applications like capacitors and high-precision sensors.
3. **Mechanical strength and toughness:**
   - The mechanical strength of a piezoelectric material determines its ability to withstand physical stresses without cracking or breaking.
   - Ceramic materials are generally brittle but have high compressive strength, while polymers like PVDF are more flexible and resilient.

4. **Electromechanical coupling factor:**
   - The electromechanical coupling factor indicates the efficiency with which a material converts electrical energy into mechanical energy and vice versa.
   - Materials with high coupling factors are ideal for applications requiring efficient energy conversion, such as ultrasonic transducers and resonators.
5. **Curie temperature:**
   - The Curie temperature is the temperature above which a piezoelectric material loses its piezoelectric properties.
   - PZT ceramics have relatively high Curie temperatures, allowing them to operate in high-temperature environments, while polymers like PVDF have lower Curie temperatures and are used in low-temperature applications.
6. **Frequency stability:**
   - The stability of piezoelectric properties over a range of frequencies is critical for applications in resonators, filters, and oscillators.
   - High-frequency stability is essential for precise timing devices, as used in communication systems and electronics.
7. **Acoustic impedance:**
   - Acoustic impedance is crucial in applications involving sound wave transmission, such as medical ultrasound imaging and underwater sonar.
   - Piezoelectric ceramics offer higher acoustic impedance, providing better coupling with dense materials like bone, whereas polymers are suited for soft tissue imaging.
8. **Flexibility and conformability:**
   - Polymeric materials such as PVDF are flexible and can conform to complex surfaces, making them suitable for wearable devices, flexible sensors, and smart textiles.
9. **Thermal stability:**
   - Thermal stability ensures that the piezoelectric properties remain consistent under varying temperatures.
   - Ceramic piezoelectrics are more stable at elevated temperatures, while polymers can degrade at higher temperatures.
10. **Poling requirements**
    - The poling process, where an electric field is applied to align the dipoles in the material, affects its piezoelectric properties.
    - Ceramic materials require a higher electric field for poling, whereas polymers like PVDF are poled at lower fields.

4.10 Properties of piezoelectric film

Piezoelectric films, such as those made from materials like PVDF, possess unique properties that make them suitable for a wide range of sensing, actuating, and energy-harvesting applications. Below are the key properties of piezoelectric films explained in detail:

1. **Piezoelectric coefficient (d_{33}, d_{31}):**
   - The piezoelectric coefficient of a film measures its ability to convert mechanical strain into electrical charge and vice versa.
   - Piezoelectric films typically have lower coefficients compared to ceramic materials like PZT, but they are flexible and can be easily conformed to various shapes.
   - The d_{31} **coefficient** is often used to describe the sensitivity of the film in applications such as pressure sensors and accelerometers.

2. **Flexibility and conformability:**
   - Piezoelectric films, particularly PVDF, are highly flexible and can be bent, stretched, or shaped without losing their piezoelectric properties.
   - This flexibility makes them ideal for applications in wearable electronics, flexible sensors, and smart textiles, where traditional rigid piezoelectric materials would not be suitable.

3. **Dielectric constant (ε_r):**
   - The dielectric constant of piezoelectric films is a measure of their ability to store electrical energy.
   - While lower than that of ceramics, the dielectric constant of PVDF films is sufficient for applications such as capacitive sensing and energy storage, especially in lightweight and portable devices.

4. **Mechanical strength and toughness:**
   - Piezoelectric films are generally more resilient to mechanical stress compared to ceramic piezoelectrics, which are brittle.
   - The films can withstand repeated bending, stretching, and compression, making them suitable for dynamic applications such as vibration sensors, impact detectors, and strain gauges.

5. **Frequency response:**
   - The frequency response of piezoelectric films is crucial in applications such as acoustic sensing and ultrasound transducers.
   - PVDF films have good high-frequency response characteristics, making them suitable for applications that require sensitivity to sound waves and vibrations.

6. **Low mass and thickness:**
   - Piezoelectric films are lightweight and can be manufactured in very thin layers, sometimes as thin as a few micrometers.
   - Their low mass is beneficial in applications where adding extra weight is undesirable, such as in aerospace, automotive, and portable electronics.

7. **Thermal stability:**
   - The thermal stability of piezoelectric films, especially PVDF, is moderate, allowing them to operate across a wide temperature range.

–   However, extreme temperatures can affect their performance. High temperatures may lead to depolarization, while very low temperatures can cause the films to become brittle.

8. **Electromechanical coupling factor:**
    –   This factor determines the efficiency with which the piezoelectric film converts electrical energy into mechanical energy (or vice versa).
    –   Although the coupling factor of piezoelectric films is lower than that of ceramics, it is still sufficient for applications like sensors and actuators, where flexibility and adaptability are more critical than high efficiency.

9. **Poling and depolarization:**
    –   The process of poling, where an electric field is applied to align the dipoles in the film, determines the strength of the piezoelectric effect.
    –   Piezoelectric films are typically poled at relatively low electric fields compared to ceramics. However, exposure to high temperatures or prolonged use can lead to depolarization, reducing their effectiveness.

10. **Chemical resistance:**
    –   Piezoelectric films, especially those made from PVDF, exhibit good resistance to chemicals, solvents, and moisture.
    –   This property makes them suitable for use in harsh environments, including outdoor applications, industrial sensors, and medical devices.

11. **Sensitivity to strain and stress:**
    –   Piezoelectric films are sensitive to mechanical deformation, allowing them to detect small changes in pressure, force, or acceleration.
    –   This high sensitivity makes them ideal for precision sensing applications such as touch sensors, strain gauges, and vibration detectors.

12. **Biocompatibility:**
    –   Certain piezoelectric films, particularly PVDF, are biocompatible, making them suitable for medical applications, such as implantable devices, biosensors, and wearable health monitors.

4.11 Smart materials featuring piezoelectric elements

Smart materials featuring piezoelectric elements are a fascinating class of materials that can convert mechanical energy into electrical energy and vice versa. This unique property allows them to be used in a wide range of applications, from sensors and actuators to energy-harvesting devices and adaptive systems. Here is a detailed exploration of these materials, their characteristics, applications, and the mechanisms behind them.

1. **Understanding piezoelectric effect:**
    –   The piezoelectric effect is the core phenomenon behind these materials. When mechanical stress is applied to a piezoelectric material, it generates an

electric charge. Conversely, applying an electric field to the material causes it to deform mechanically. This bidirectional capability allows piezoelectric materials to act as both sensors (detecting changes in force or pressure) and actuators (producing movement or vibration).

2. **Types of piezoelectric materials:**
   - **Crystalline materials:** Quartz is a classic example of a naturally occurring piezoelectric crystal. However, its application is limited due to its low piezoelectric coefficient.
   - **Ceramic materials:** PZT is the most widely used piezoelectric ceramic due to its strong piezoelectric properties. These materials are engineered to achieve high sensitivity and robustness.
   - **Polymeric materials:** PVDF and its copolymers are piezoelectric polymers known for their flexibility, lightweight, and ease of processing. These are particularly useful in flexible sensors and wearable devices.

3. **Properties of piezoelectric smart materials:**
   - **Direct piezoelectric effect:** This effect is the generation of an electric charge in response to mechanical stress. This property is used in sensors, such as accelerometers, pressure sensors, and vibration detectors.
   - **Inverse piezoelectric effect:** This effect is the mechanical deformation of the material when an electric field is applied. This is exploited in actuators, like precision positioning systems and ultrasonic transducers.
   - **High sensitivity and precision:** Piezoelectric materials can detect even slight changes in force or pressure, making them ideal for precision measurement applications.
   - **Energy-harvesting capabilities:** Piezoelectric materials can convert ambient mechanical energy, like vibrations or footsteps, into electrical energy, which can be stored and used to power small devices.

4. **Applications of piezoelectric smart materials:**
   - **Sensors:**
     - Piezoelectric materials are widely used in sensors due to their high sensitivity. Examples include:
       - **Accelerometers:** Used in aerospace, automotive, and consumer electronics for measuring acceleration, tilt, and vibration.
       - **Pressure sensors:** Found in industrial systems, medical devices, and consumer electronics for monitoring pressure changes.
       - **Touch sensors:** Integrated into devices for responsive touch interfaces.
   - **Actuators:**
     - Piezoelectric actuators are employed in applications where precise movement control is required:
       - **Ultrasonic transducers:** Used in medical imaging (e.g., ultrasound devices), nondestructive testing, and sonar systems.

- **Precision positioning:** Employed in applications such as optical alignment, micromanipulation, and nanoscale motion control in scientific instruments.
- **Adaptive optics:** In telescopes and cameras, piezoelectric actuators adjust mirrors and lenses to correct for optical aberrations.

- **Energy harvesting:**
  - Piezoelectric materials are effective in harvesting energy from mechanical vibrations or movements:
    - **Vibration energy harvesters:** They convert environmental vibrations (like those from machines or infrastructure) into electrical energy for powering low-power devices.
    - **Wearable energy harvesters:** They capture energy from body movements (e.g., walking, running) to power small electronics such as sensors, health monitors, or communication devices.

- **Medical devices:**
  - Piezoelectric materials are used in a range of medical applications:
    - **Ultrasonic imaging:** This provides high-resolution imaging for diagnostics.
    - **Drug delivery systems:** Piezoelectric pumps control the precise delivery of medications in implantable devices.
    - **Wearable health monitors:** Flexible piezoelectric films measure physiological parameters such as heartbeat, respiration, and muscle activity.

- **Structural health monitoring:**
  - Embedded piezoelectric sensors monitor the integrity of critical structures, such as bridges, buildings, and aircraft. They detect early signs of stress, cracks, or other damages.

5. **Smart systems featuring piezoelectric elements:**
   - **Self-adaptive systems:** Smart materials with piezoelectric components can adapt to changing environmental conditions in real time. For instance, they can adjust the stiffness of a structure dynamically in response to external loads.
   - **Feedback control systems:** In advanced robotics and precision manufacturing, piezoelectric elements enable precise feedback and control, ensuring accurate and repeatable operations.

6. **Challenges and future directions:**
   - **Material optimization:** Researchers are continuously working to improve the efficiency, durability, and response speed of piezoelectric materials, especially in polymeric forms.
   - **Integration with other technologies:** Combining piezoelectric materials with other smart materials, such as shape-memory alloys or magnetostrictive materials, can create hybrid systems with enhanced functionality.

- **Scalability and cost-effectiveness:** Making piezoelectric materials more affordable and easier to mass-produce will broaden their use in consumer electronics, IoT devices, and beyond.

One practical example of piezoelectric smart materials can be seen in energy-harvesting floor tiles, a sustainable innovation increasingly used in public spaces, train stations, and sports arenas.

How it works

The floor tiles are embedded with piezoelectric elements, typically made from materials like PZT or PVDF. When pressure is applied to these tiles – such as when someone walks across them – the mechanical stress triggers the piezoelectric effect. This effect causes the materials to generate an electric charge due to the deformation of their crystalline structure. The generated charge is collected and stored in batteries or directly used to power nearby devices.

Components and mechanism

- **Piezoelectric materials:** The core of the technology is the piezoelectric elements, which are strategically placed beneath the surface of the tile. When the material experiences compressive or tensile stress, it produces an electrical charge proportional to the applied force.
- **Energy storage system:** The electrical energy generated is either stored in a battery or capacitor, depending on the application. Advanced systems can store and regulate the power for later use.
- **Power management circuit:** This circuit converts the piezoelectric-generated voltage into a stable form suitable for powering devices such as LED lights or sensors.
- **Tile structure:** The tiles themselves are designed to be robust, durable, and capable of withstanding constant pressure while maintaining efficiency in energy generation.

Applications

1. **Public spaces and sports arenas:** Energy-harvesting tiles are increasingly used in high-traffic areas such as airports, shopping malls, and stadiums. The foot traffic of thousands of people daily generates significant energy, which is used to power lighting systems or electronic displays. For instance, at the Paris Marathon, energy-harvesting tiles were installed along the race path, collecting energy from runners' footsteps to power nearby lighting systems.
2. **Sustainable architecture:** In green building projects, these tiles contribute to energy efficiency by generating renewable electricity. In smart cities, such technolo-

gies are integrated into the infrastructure to reduce reliance on traditional energy sources and lower carbon footprints.

3. **Transportation hubs:** Train stations or airports use piezoelectric tiles in walkways or entry points to generate electricity that powers information displays, signage, or even charging stations. The more people pass through, the more energy is harvested.

Advantages

- **Sustainability:** Piezoelectric tiles provide a clean energy source without emitting greenhouse gases or relying on fossil fuels. They make use of kinetic energy that would otherwise be wasted.
- **Scalability:** The technology can be scaled from small installations in homes or offices to large commercial or urban environments, offering versatility.
- **Low maintenance:** Once installed, the system requires minimal maintenance, making it cost-effective over time.

Challenges

- **Efficiency:** The amount of energy generated from footsteps alone is relatively low, which means large installations are needed to produce significant power.
- **Cost:** Initial installation costs can be high, although advances in materials and mass production are reducing expenses over time.
- **Durability:** The tiles must be robust enough to withstand constant pressure and environmental factors such as moisture and temperature changes without degrading.

Future directions

- **Integration with IoT:** Future energy-harvesting systems could be linked with smart city infrastructures, allowing data to be collected from the tiles to monitor foot traffic, adjust lighting automatically, or optimize energy storage systems.
- **Hybrid systems:** Combining piezoelectric tiles with solar panels or other renewable energy sources can further enhance the overall energy output and efficiency.

4.12 Smart composite laminate with embedded piezoelectric actuators

Smart composite laminate with embedded piezoelectric actuators is an advanced material system that integrates piezoelectric actuators within composite laminates, allowing for active control of structural properties such as vibration damping, shape control, and health monitoring [31]. These smart laminates are increasingly used in aerospace, automotive, and civil engineering for their ability to adapt to changing conditions and improve performance.

How it works

- **Composite structure:** The laminate consists of multiple layers of composite materials (e.g., carbon fiber and glass fiber) bonded together with a polymer matrix (e.g., epoxy resin). These composites provide high strength-to-weight ratios, making them ideal for lightweight structural applications.
- **Embedded piezoelectric actuators:** Piezoelectric actuators, typically made from materials like PZT, are embedded between the composite layers. These actuators can convert electrical energy into mechanical strain and vice versa. When a voltage is applied to the piezoelectric actuators, they generate strain, allowing precise control of the laminate's deformation.

Working principle

1. **Electrical input:** An external controller applies an electrical signal to the embedded piezoelectric actuators. The voltage induces mechanical strain in the actuators.
2. **Actuation and control:** The strain generated by the actuators is transferred to the surrounding composite layers, resulting in controlled deformation of the laminate. This mechanism can be used to adjust the shape of the structure, suppress vibrations, or apply active control in real time.
3. **Sensing and feedback:** In addition to actuation, the piezoelectric elements can act as sensors, detecting structural changes, vibrations, or damage by measuring the generated electrical charge when the structure undergoes deformation. This data is used for real-time monitoring and feedback control.

Applications

1. **Aerospace:** In aircraft wings and helicopter rotor blades, smart composite laminates with embedded piezoelectric actuators enable active shape control, reducing drag and improving aerodynamic efficiency. They can also be used in morphing wings that change the shape in response to flight conditions.
2. **Vibration control:** In mechanical systems and structures, these smart laminates actively dampen unwanted vibrations. For example, in precision machinery, the actuators can counteract vibrations that would otherwise reduce accuracy.
3. **Structural health monitoring:** The piezoelectric sensors within the laminate detect damage or fatigue in real time. This self-monitoring capability is valuable in critical structures such as bridges or aircraft, where early detection of damage can prevent catastrophic failures.

4. **Robotics:** In adaptive robotic structures, piezoelectric actuators are used to create flexible, responsive components that can change the shape or stiffness on demand, improving maneuverability and functionality.

Advantages
- **Lightweight and high strength:** The composite structure maintains high performance while adding minimal weight, crucial for aerospace and automotive applications.
- **Real-time control:** Embedded actuators enable fast response to changes in load, temperature, or external forces, improving performance and safety.
- **Multifunctionality:** The combination of actuation, sensing, and structural reinforcement in a single material system reduces the need for additional components and simplifies the design.

Challenges
- **Complex manufacturing:** Integrating piezoelectric elements within composite laminates requires precision and advanced manufacturing techniques.
- **Cost:** The cost of piezoelectric materials and complex manufacturing processes can be high, limiting widespread adoption.
- **Durability:** Ensuring the long-term reliability and performance of embedded actuators in harsh environments remains a challenge.

A practical example of smart composite laminates with embedded piezoelectric actuators is found in aerospace wing morphing technology.

Application in morphing aircraft wings:
In modern aircraft, there is a need for wings that can dynamically change the shape during flight to improve performance, reduce drag, and optimize fuel efficiency. Traditional control surfaces, such as ailerons and flaps, introduce additional drag and have limited flexibility. Smart composite laminates with embedded piezoelectric actuators provide a solution by enabling smooth, real-time wing morphing.

How it works
- **Embedded piezoelectric actuators:** In this system, piezoelectric actuators are embedded within the composite layers of the aircraft wing. When a voltage is applied, these actuators create strain, causing the wing's shape to change subtly. Unlike mechanical hinges or joints, these actuators allow for continuous, smooth deformation.
- **Active shape control:** The smart laminate can adjust the wing's camber (curvature) or twist angle during different flight phases, such as take-off, cruising, or

landing. This adaptability improves aerodynamic efficiency and can lead to significant fuel savings.
- **Integrated sensing:** The same piezoelectric elements can act as sensors, detecting aerodynamic loads, wing vibrations, or potential structural issues. The data is fed back to the control system, allowing for precise adjustments and real-time monitoring.

Benefits
- **Fuel efficiency:** By optimizing wing shape for different flight conditions, morphing wings reduce drag and increase fuel efficiency, which is critical in both commercial aviation and military applications.
- **Enhanced maneuverability:** Aircraft with morphing wings can achieve better control and performance, particularly in military applications where agility is essential.
- **Reduced maintenance:** Unlike mechanical control surfaces, the embedded actuators have fewer moving parts, leading to lower maintenance requirements and longer service life.

Example in action
The NASA "Mission Adaptive Wing" project and subsequent research have explored using smart composite laminates with embedded piezoelectric actuators to develop wings that can adapt during flight. In experimental aircraft, this technology has shown potential in reducing drag by up to 20%, significantly improving the efficiency.

4.13 SAW filters

Surface acoustic wave (SAW) filters are devices used in signal processing to filter specific frequencies in radio frequency (RF) signals. They are commonly used in communication systems such as mobile phones, satellite receivers, and wireless networks [32]. SAW filters take an advantage of SAWs – mechanical waves that travel along the surface of a material – to perform this filtering function.

How SAW filters work
1. **Piezoelectric substrate:** SAW filters are typically made from piezoelectric materials such as quartz, lithium niobate, and lithium tantalate. The piezoelectric effect allows the conversion between electrical signals and mechanical (acoustic) waves.

2. **Interdigital transducers (IDTs):** The filter consists of two main transducers: an input IDT and an output IDT. When an RF signal is applied to the input IDT, it generates acoustic waves on the surface of the substrate.
3. **SAW propagation:** The acoustic waves travel along the surface of the substrate. As they propagate, they are subject to interference patterns that help filter specific frequencies.
4. **Output transducer:** The output IDT reconverts the acoustic waves back into an electrical signal. Only the desired frequency range is transmitted through, while other frequencies are attenuated, achieving the filtering effect.

Applications of SAW filters

-   **Mobile communication:** SAW filters are extensively used in mobile phones to filter out unwanted frequencies and to separate different communication bands.
-   **Television broadcasting:** These filters help in selecting specific channels by filtering out unnecessary signals.
-   **GPS systems:** SAW filters are used to ensure only the correct GPS signals are processed, enhancing accuracy.
-   **Radar systems:** In radar applications, SAW filters are used to refine the signal and eliminate noise.

Advantages

-   **High selectivity:** SAW filters provide precise control over frequency ranges, allowing them to target specific frequencies very effectively.
-   **Compact size:** Their small size makes them ideal for integration into mobile and portable devices.
-   **Low cost:** Mass production of SAW filters is relatively inexpensive, making them widely accessible for consumer electronics.

Limitations

-   **Temperature sensitivity:** SAW filters can be affected by temperature variations, which may alter their performance.
-   **Frequency range:** While effective for RF and microwave frequencies, SAW filters have limitations when operating at very high frequencies, making them unsuitable for certain applications.

Practical example

Practical example of SAW filters in smartphones

In modern smartphones, multiple communication standards are in use, including GSM, 3G, 4G LTE, and 5G, along with Wi-Fi and Bluetooth. Each of these standards operates at different frequency bands, and to ensure seamless communication without interference, precise filtering is required [32]. This is where SAW filters come into play.

How SAW filters work in smartphones

1. **Signal reception and transmission:** When your smartphone connects to a cellular network, it needs to transmit and receive signals on specific frequency bands. These bands are often crowded with signals from different sources, so filtering is essential to isolate the correct frequencies for communication.
2. **Filtering specific frequency bands:** SAW filters are integrated into the smartphone's RF front-end module. They are designed to allow only the desired frequency range to pass through while attenuating or blocking signals outside of this range. For example, a SAW filter can target the 900 MHz band for GSM communication while filtering out signals from nearby frequency bands that could cause interference.
3. **IDTs:** The core of a SAW filter is IDTs, which convert the electrical signals from the phone's antenna into SAWs on a piezoelectric substrate (e.g., quartz or lithium niobate). As these waves propagate, they interact in a way that selectively amplifies or suppresses specific frequencies.
4. **Reconversion to electrical signals:** After the filtering process, the acoustic waves are converted back into electrical signals by the output transducer, which are then processed by the smartphone's radio transceiver.

Practical application in smartphone communication

Let us consider a scenario where you are using 4G LTE data on your smartphone. The LTE standard operates on different frequency bands depending on the region. For example, Band 20 operates in the 800 MHz range. The SAW filter in your phone will ensure that only signals in the 800 MHz range are processed, while signals from other nearby bands, like those for Wi-Fi (2.4 GHz) or GSM (900 MHz), are filtered out.

This filtering is crucial because it allows your smartphone to maintain a stable and high-quality connection without interference from other signals. It also prevents the phone's antenna from picking up unwanted noise, which could degrade the call quality, data speeds, and battery life.

Key advantages in smartphones

- **Compact design:** SAW filters are small and lightweight, making them ideal for space-constrained devices like smartphones.
- **Cost-effective:** SAW filters are inexpensive to manufacture, contributing to the overall affordability of mobile devices.

- **High precision:** These filters provide accurate control over the frequency bands, which is essential for handling multiple communication standards in a single device.

Real-world example

In a typical 4G LTE smartphone, multiple SAW filters are used to handle different frequency bands. For instance, a phone supporting both LTE Band 3 (1,800 MHz) and LTE Band 7 (2,600 MHz) will have separate SAW filters for each band, ensuring clear and interference-free communication in both bands [33]. Without these filters, the phone would struggle to differentiate between the signals from each band, leading to poor performance and dropped connections.

Impact on user experience

- **Better call quality:** SAW filters reduce noise and interference, resulting in clearer voice calls.
- **Improved data transmission:** By filtering out irrelevant frequencies, these filters enhance data speeds and ensure stable connections.
- **Energy efficiency:** By preventing the phone from processing unnecessary signals, SAW filters contribute to longer battery life.

SAW filters play a critical role in smartphones by ensuring that they can reliably and efficiently manage multiple communication standards without interference. This capability is key to delivering the seamless and high-quality connectivity users expect from modern mobile devices.

4.14 Summary

This chapter explores various phenomena and applications of materials that exhibit piezoelectric, electrostrictive, and pyroelectric properties. Piezoelectric materials, such as PZT and PVDF films, generate an electric charge under mechanical stress, enabling their use in sensors, actuators, and energy-harvesting devices. Electrostriction, which involves the deformation of dielectric materials in response to an electric field, complements piezoelectric effects in precision applications. Pyroelectricity, the generation of charge in response to temperature changes, further extends these materials' utility. Industrial piezoelectric materials are classified based on their properties, such as high sensitivity and stability, with PZT being widely used in actuators and PVDF in flexible sensors. The properties of commercial piezoelectric materials are essential for designing smart devices, from medical diagnostics to wearable tech. Additionally, smart composite laminates with embedded piezoelectric actuators allow for structural health monitoring and adaptive controls. SAW filters, which leverage piezoelectric effects, are vital in communication systems for filtering specific frequencies. Overall, the chapter covers the principles, properties, and diverse applications of piezoelectric smart materials, emphasizing their integration into advanced technological systems.

Review questions

1. What are piezoelectric smart materials?
2. Explain the mechanism behind piezoelectricity.
3. What is electrostriction, and how does it differ from piezoelectricity?
4. Define pyroelectricity and describe its applications.
5. What are some common industrial applications of piezoelectric materials?
6. Describe the structure and properties of PZT film.
7. What are the key characteristics of PVDF film in piezoelectric applications?
8. How does PVDF differ from PZT in terms of flexibility and applications?
9. List the main properties of commercial piezoelectric materials.
10. How does temperature affect the performance of piezoelectric materials?
11. Explain the importance of the dielectric constant in piezoelectric materials.
12. What are some advantages of using piezoelectric films in sensors?
13. Discuss the role of piezoelectric elements in smart materials.
14. What are smart composite laminates, and how are they used?
15. How do embedded piezoelectric actuators function in composite materials?
16. What is the purpose of SAW filters in communication systems?
17. Explain the working principle of a SAW filter.
18. How do piezoelectric materials contribute to structural health monitoring?
19. What are some challenges in using piezoelectric materials in harsh environments?
20. Discuss the difference between direct and converse piezoelectric effects.
21. What are the primary applications of electrostrictive materials?
22. How is pyroelectricity utilized in temperature sensing?
23. What materials are commonly used for pyroelectric applications?
24. Describe how piezoelectric materials are used in energy harvesting.
25. What are the key design parameters for creating efficient piezoelectric devices?
26. How do piezoelectric materials help in vibration control?
27. Explain the process of polarization in piezoelectric materials.
28. How do mechanical stress and electric fields interact in piezoelectric materials?
29. What are the benefits of using PVDF film in wearable technology?
30. Describe a practical application of PZT film in the medical field.
31. What role do piezoelectric materials play in MEMS?
32. How does the thickness of a piezoelectric film affect its performance?
33. What is the significance of piezoelectric coefficients in material selection?
34. How do piezoelectric sensors work in automotive applications?

35. What are the advantages of using SAW filters over other types of filters?
36. Explain how smart materials featuring piezoelectric elements are used in adaptive systems.
37. What are some common challenges in designing smart materials with embedded piezoelectric actuators?
38. Describe the use of piezoelectric materials in noise-cancellation systems.
39. How do piezoelectric films contribute to high-frequency signal processing?
40. What are the key factors affecting the performance of smart composite laminates?
41. Explain how piezoelectric actuators can be used in precision positioning systems.
42. How do temperature and humidity affect the behavior of piezoelectric materials?
43. What is the role of SAW filters in mobile communication technology?
44. Discuss how piezoelectricity is applied in ultrasound imaging.
45. What are some emerging applications of piezoelectric materials in consumer electronics?
46. How is the piezoelectric effect utilized in musical instruments?
47. What are the environmental considerations when using piezoelectric materials?
48. Explain how piezoelectric elements are integrated into aerospace applications.
49. What are the different types of piezoelectric sensors and their applications?
50. How do smart materials with piezoelectric elements enhance the performance of robotics?

5 Shape-memory (alloys) smart materials

5.1 Introduction

Shape-memory alloys (SMAs) are advanced smart materials with the unique capability to revert to a predetermined shape after being deformed, triggered by changes in temperature or stress [34]. This behavior arises from a reversible phase transformation between two crystalline phases: martensite (low-temperature phase) and austenite (high-temperature phase). When cooled, SMAs enter the martensitic phase, allowing them to be easily deformed. Upon heating, the alloy transitions to the austenitic phase, causing it to return to its original shape – a phenomenon known as the shape-memory effect (SME) [35].

In addition to SME, SMAs also exhibit superelasticity (or pseudoelasticity), where the material can undergo large strains when stressed and recover its original shape immediately upon unloading, without needing a temperature change. This occurs due to stress-induced phase transformation between the austenite and martensite phases.

Common SMAs include nickel–titanium (NiTi) alloys, also known as nitinol, which are widely used in medical devices such as stents and orthodontic wires, as well as in actuators, sensors, and robotics. The unique properties of SMAs make them ideal for applications requiring precise actuation, adaptive structures, and energy absorption [5]. Their ability to adapt to environmental changes and recover from deformation gives SMAs a prominent role in the development of intelligent systems, enabling innovation in fields such as aerospace, automotive, and biomedical engineering.

5.2 Background on shape-memory alloys (SMA)

SMAs have their roots in the discovery of SME in the early twentieth century. The phenomenon was first observed in 1932 in a gold–cadmium alloy, but it was not until the 1960s that SMAs gained significant attention with the discovery of the nickel–titanium alloy (nitinol) by the Naval Ordnance Laboratory in the United States. Nitinol, which stands for nickel–titanium, displayed both shape memory and superelastic properties, sparking interest in its potential applications.

The development of SMAs was driven by the need for materials that could perform specific functions autonomously without requiring complex mechanical systems. Over time, SMAs have evolved from being a scientific curiosity to becoming integral in diverse fields such as aerospace, biomedical engineering, and robotics.

The underlying mechanism of SMAs is the reversible phase transformation between martensite (low-temperature phase) and austenite (high-temperature phase).

https://doi.org/10.1515/9783111379623-005

This transformation allows SMAs to recover their original shape after deformation, either through heating (SME) or unloading (superelasticity).

Since the initial discovery of nitinol, extensive research has been conducted to develop a wide range of SMAs with tailored properties, including copper-based alloys, iron-based alloys, and various nickel–titanium composites. The ability of SMAs to "remember" shapes and their durability in cyclic applications make them ideal for both high-tech and everyday applications, from deployable space structures to eyeglass frames and medical stents.

The field of SMAs continues to expand as new alloys and manufacturing techniques emerge, enabling more sophisticated and efficient smart systems that leverage the adaptive and transformative characteristics of these materials.

5.3 Nickel alloys

Nickel alloys are a group of metal alloys primarily composed of nickel, often combined with other elements such as chromium, iron, copper, and molybdenum. These alloys are valued for their unique properties, including corrosion resistance, high-temperature stability, and magnetic characteristics. Here is a brief overview of nickel alloys, focusing on their types, properties, and applications.

Types of nickel alloys

1. **Nickel–copper alloys (e.g., Monel)**
    -   **Composition**: Typically 60–70% nickel and 30–40% copper, with small amounts of iron, manganese, and other elements.
    -   **Properties**: High resistance to corrosion, excellent strength, and toughness. Monel alloys are known for their resistance to seawater and acidic environments.
    -   **Applications**: Marine engineering, chemical processing, and aerospace components.
2. **Nickel–chromium alloys (e.g., Inconel)**
    -   **Composition**: Nickel (usually 60–70%) combined with chromium (15–25%) and often iron, molybdenum, or other elements.
    -   **Properties**: Exceptional resistance to oxidation and corrosion at high temperatures. Inconel alloys maintain their strength and stability in extreme conditions.
    -   **Applications**: Aerospace, gas turbines, nuclear reactors, and chemical processing equipment.

3. **Nickel–iron alloys (e.g., permalloy)**
   - **Composition**: Predominantly nickel (around 80%) with iron (about 20%) and sometimes small additions of other elements.
   - **Properties**: High magnetic permeability, low coercivity, and excellent magnetic shielding properties.
   - **Applications**: Magnetic cores, transformers, inductors, and magnetic shielding.
4. **Nickel–molybdenum alloys**
   - **Composition**: Nickel combined with molybdenum and sometimes small amounts of chromium and other elements.
   - **Properties**: High strength, excellent corrosion resistance, and good performance at high temperatures.
   - **Applications**: Chemical processing, petrochemical industry, and high-stress applications.

Properties

- **Corrosion resistance**: Nickel alloys exhibit high resistance to corrosion and oxidation, making them suitable for harsh environments.
- **High-temperature stability**: Many nickel alloys retain their strength and stability at elevated temperatures, which is crucial for high-temperature applications.
- **Magnetic properties**: Nickel alloys, particularly those with high nickel content, exhibit specific magnetic properties such as high permeability, making them useful in electronic and magnetic applications.
- **Mechanical strength**: Nickel alloys are known for their strength and durability, often making them suitable for demanding applications requiring high mechanical performance.

Applications

- **Aerospace**: Nickel alloys are used in aerospace components like turbine blades and exhaust systems due to their high-temperature strength and corrosion resistance.
- **Chemical processing**: Alloys such as Monel and Inconel are used in chemical processing equipment for their resistance to corrosive chemicals.
- **Electronics**: Nickel–iron alloys like Permalloy are used in magnetic cores and transformers due to their excellent magnetic properties.
- **Marine engineering**: Nickel–copper alloys are employed in marine environments for their resistance to seawater corrosion.

Nickel alloys are a versatile group of materials known for their strength, corrosion resistance, and specialized properties. They are essential in various industries, including aerospace, chemical processing, electronics, and marine engineering. By selecting the appropriate nickel alloy, engineers and designers can achieve optimal performance for specific applications, making these alloys integral to modern technology and infrastructure.

Nickel alloys are widely used in practical applications across various industries due to their exceptional properties. For example, Inconel alloys, with their high-temperature stability and corrosion resistance, are commonly used in the aerospace industry for turbine blades and exhaust systems, where they endure extreme conditions and maintain performance. Monel alloys, known for their resistance to seawater corrosion, are used in marine applications like shipbuilding and offshore oil rigs to ensure durability and longevity in harsh environments. Permalloy is utilized in electronic devices such as transformers and magnetic shielding materials due to its excellent magnetic permeability, which enhances the efficiency of electronic components. These practical applications highlight the versatility and importance of nickel alloys in modern technology and industry.

5.4 Titanium alloy (nitinol)

Titanium alloy, specifically nitinol, is a notable SMA composed primarily of nickel and titanium. It is renowned for its unique combination of shape-memory and superelastic properties. Here is a brief overview:

Composition

- **Nickel–titanium (nitinol)**: Typically consists of about 50–60% nickel and 40–50% titanium. The exact ratio can be adjusted to tailor specific properties.

Properties

- **SME**: Nitinol can return to its predetermined shape when heated above a certain transition temperature. This happens due to a phase transformation between the martensite (low-temperature) and austenite (high-temperature) phases.
- **Superelasticity**: Nitinol exhibits high flexibility and can undergo significant deformation under stress and return to its original shape when the stress is removed. This property is due to stress-induced phase transformations.
- **Biocompatibility**: Nitinol is highly biocompatible, making it ideal for medical implants and devices.

Practical example

In the **medical field**, nitinol is used for **stents**, which are small mesh tubes inserted into blood vessels to keep them open. The stents are deployed in their collapsed form and expand to their original shape when they reach the body temperature, providing effective treatment with minimal invasiveness.

5.5 Material characteristics of nitinol

Nitinol, a nickel–titanium alloy, possesses several unique material characteristics that make it valuable for a wide range of applications, particularly in medical devices and engineering [36]. Below are the key material characteristics of nitinol:

1. SME
   - **Description**: Nitinol can "remember" and return to its original shape after being deformed when heated above a certain transition temperature. This occurs due to a reversible phase transformation between its martensitic (low-temperature) and austenitic (high-temperature) phases.
   - **Application**: It is used in medical stents, eyeglass frames, and actuators.
2. Superelasticity (pseudoelasticity)
   - **Description**: Nitinol exhibits high elasticity, allowing it to undergo significant deformation (up to 10%) and return to its original shape without permanent deformation when the load is removed. This behavior is stress-induced and happens at temperatures above the alloy's transformation point.
   - **Application**: It is utilized in orthodontic archwires, flexible surgical tools, and fatigue-resistant applications.
3. Biocompatibility
   - **Description**: Nitinol is highly biocompatible and does not react adversely with body tissues, making it suitable for long-term medical implants.
   - **Application**: It is commonly used in medical implants, such as vascular stents and guidewires.
4. Corrosion resistance
   - **Description**: Nitinol has good corrosion resistance, particularly in biological environments, which is critical for its use in medical devices.
   - **Application**: It ensures long-term reliability in implants and devices exposed to bodily fluids.

5. Fatigue resistance
   – **Description**: Nitinol is highly resistant to fatigue, which means it can undergo repeated cycles of deformation and recovery without significant degradation.
   – **Application**: It is used in applications requiring long-term performance, such as cardiovascular stents and actuators.
6. Ductility and toughness
   – **Description**: Nitinol is highly ductile and can be easily formed into complex shapes while retaining its mechanical properties. It also has excellent toughness, making it resistant to fracture under stress.
   – **Application**: It allows for the manufacturing of intricate medical devices, such as wire mesh filters and flexible surgical instruments.
7. Thermal stability
   – **Description**: Nitinol's phase transformation temperatures are stable, allowing for precise control over its shape-memory and superelastic behaviors based on temperature.
   – **Application**: It is critical for temperature-sensitive applications such as actuators and sensors.

A practical example of nitinol is its use in **vascular stents.** Nitinol stents are small mesh tubes implanted in arteries to keep them open and restore proper blood flow. These stents are compressed into a small shape for insertion, and once inside the blood vessel, they expand back to their original shape due to the SME triggered by the body temperature. The superelasticity of nitinol allows the stent to flex and move with the artery, maintaining its functionality over time without causing damage or discomfort to the patient. This characteristic makes nitinol stents highly effective in treating cardiovascular conditions such as blockages or narrowing of arteries.

5.6 Martensitic transformations

Martensitic transformations refer to a solid-state phase change that occurs without diffusion, typically in response to changes in temperature or applied stress [37]. These transformations are critical in SMAs like nitinol and are responsible for their unique properties, such as SME and superelasticity. Below is an explanation of martensitic transformations:

Key aspects of martensitic transformations

1. **Phase change without diffusion**:
   - Martensitic transformation involves a coordinated movement of atoms that changes the crystal structure of the material without the need for atom diffusion. This makes the transformation very rapid.
2. **Crystal structure change**:
   - The transformation typically occurs between the austenite phase (high-temperature phase) and the martensite phase (low-temperature phase). Austenite has a more ordered, stable crystal structure, while martensite is a less stable, distorted phase.
3. **Temperature-dependent transformation**:
   - The transformation between austenite and martensite phases happens at specific temperatures:
     - **Martensite start temperature (M_s)**: The temperature at which martensite begins to form upon cooling.
     - **Martensite finish temperature (M_f)**: The temperature at which the transformation to martensite is complete.
     - **Austenite start temperature (A_s)**: The temperature at which austenite begins to form upon heating.
     - **Austenite finish temperature (A_f)**: The temperature at which the transformation back to austenite is complete.
4. **Reversibility**:
   - The martensitic transformation is reversible in SMAs. Upon cooling, the material transforms to martensite, and upon heating, it reverts to austenite, allowing the material to "remember" its original shape.
5. **Stress-induced transformation**:
   - In addition to temperature, applied stress can also induce martensitic transformation. This is the basis for superelasticity in SMAs, where the material deforms under stress and then recovers its shape once the stress is removed.

Practical example

A practical example of martensitic transformations is seen in **eyeglass frames made of nitinol**. These frames can be bent, twisted, or deformed significantly without breaking. When the deformation occurs, the material enters its martensitic phase, allowing it to flex. Once the force is removed, the material undergoes a transformation back to its original austenitic phase and returns to its original shape. This property makes nitinol ideal for durable, flexible eyeglass frames that can withstand rough handling while maintaining their structure.

5.7 Austenitic formations

Austenitic transformations refer to the phase change in materials, particularly in SMAs, where the crystal structure transforms from martensite to austenite [38]. This transformation is crucial to the functionality of SMAs, enabling properties such as SME and superelasticity.

Key aspects of austenitic transformations

1. **High-temperature phase**:
   – The austenite phase is stable at higher temperatures and has a more ordered, cubic crystal structure compared to martensite. This phase is responsible for the material "remembering" its original shape.
2. **Temperature-dependent transformation**:
   – The transformation occurs as the material is heated:
      – **Austenite start temperature (A_s)**: The temperature at which austenite begins to form as the material is heated.
      – **Austenite finish temperature (A_f)**: The temperature at which the transformation to austenite is complete.
3. **SME**:
   – In SMAs, when the material is deformed in its martensitic phase (low-temperature phase), heating it above the austenite finish temperature (A_f) causes it to revert to the austenitic phase and return to its original, predeformed shape.
4. **Reversibility**:
   – The transition between martensite and austenite is fully reversible, allowing for repeated cycles of shape change and recovery, a key feature in smart materials.

Practical example

In **nitinol stents** used in medical procedures, the austenitic transformation plays a vital role. When the stent is deployed in the body, the body's temperature triggers the transition to the austenitic phase, causing the stent to expand and restore the blood flow by opening the blocked artery. The ability to transition between phases ensures that the stent can be easily inserted and then activated at the right temperature to function effectively.

5.8 Thermoelastic martensitic transformations

Thermoelastic martensitic transformations refer to a specific type of phase transformation in SMAs, where the change between the martensite and austenite phases is both temperature-dependent and reversible [39]. These transformations are fundamental to the behavior of SMAs, enabling properties such as SME and superelasticity.

Key aspects of thermoelastic martensitic transformations

1. **Reversible transformation**:
   – The transformation between the martensite (low-temperature) and austenite (high-temperature) phases is reversible. When the temperature changes, the material can switch phases without permanent deformation.
2. **Temperature control**:
   – The transformation is triggered by specific temperature ranges:
     – **Cooling** induces the formation of martensite from austenite.
     – **Heating** causes the reverse transformation from martensite back to austenite.
3. **Elastic deformation**:
   – The transformation is "thermoelastic" because the martensite phase deforms elastically. This allows the material to revert to its original shape when the temperature changes, a key feature of SME.
4. **Low hysteresis**:
   – The transformation occurs with low energy loss, meaning there is minimal temperature difference between the start and finish of the phase changes. This makes the transformation efficient and highly controllable.

Practical example

A common example is in **actuators made from nitinol** (a nickel–titanium alloy). These actuators can change the shape when heated, performing tasks like opening a valve or shifting a component. Once cooled, they return to their original shape. The thermoelastic nature of the transformation allows for precise, repeatable motion, making these actuators useful in applications ranging from robotics to medical devices.

5.9 Cu-based SMA

Cu-based SMAs are a category of SMAs primarily composed of copper along with other elements such as aluminum, zinc, or nickel [40]. These alloys exhibit the SME and superelasticity, similar to more commonly known SMAs like nitinol.

Key aspects of Cu-based SMAs

1. **Composition**:
   - Common Cu-based SMAs include:
     - **Cu–Al–Ni**: Copper–aluminum–nickel
     - **Cu–Zn–Al**: Copper–zinc–aluminum
2. **Advantages**:
   - **Cost-effective**: Cu-based SMAs are generally less expensive than NiTi (nitinol) alloys.
   - **High transformation temperatures**: These alloys can have higher transformation temperatures, making them suitable for applications in elevated temperature environments.
   - **Good damping properties**: They offer effective vibration damping in various mechanical systems.
3. **Limitations**:
   - **Brittleness**: Cu-based SMAs are more brittle compared to nitinol, limiting their use in applications requiring extensive mechanical deformation.
   - **Lower fatigue resistance**: They may have lower fatigue life, which affects their performance under repeated cycles.

Practical example in brief

A practical example of Cu–Al–Ni alloys in aerospace applications is their use in **actuators** for controlling aerodynamic surfaces. These actuators take advantage of the high transformation temperature and SME of Cu–Al–Ni SMAs. When exposed to temperature changes, the alloy changes the shape, enabling precise adjustments of surfaces like wing flaps or airfoil positions. This helps optimize flight performance in varying conditions. The reliability and ability to function in high-temperature environments make Cu–Al–Ni actuators ideal for such aerospace tasks.

5.10 Chiral materials

Chiral materials are substances with structures that exhibit chirality, meaning they lack mirror symmetry [41]. In other words, they cannot be superimposed on their mirror images. This property leads to unique optical and mechanical behaviors that are exploited in various applications.

Key aspects of chiral materials

1. **Chirality**:
   - **Definition**: Chirality refers to a property where an object or a structure is not superimposable on its mirror image. In materials, this means that the arrangement of atoms or molecules results in a structure that cannot be aligned with its mirror image.
   - **Types**: Chiral materials can be molecular (e.g., chiral molecules), structural (e.g., chiral lattices), or macroscopic (e.g., chiral metamaterials).
2. **Optical properties**:
   - **Optical activity**: Chiral materials can rotate the plane of polarized light due to their chiral nature, which affects how light interacts with the material.
   - **Circular dichroism**: Chiral materials exhibit different absorption rates for left-handed and right-handed circularly polarized light, which is useful in identifying molecular structures.
3. **Mechanical properties**:
   - **Chiral elasticity**: Chiral materials can exhibit unique mechanical behaviors, such as unusual deformation patterns, due to their non-mirror symmetric structures.
   - **Applications**: These properties are exploited in designing materials with specific mechanical responses, such as flexible or self-healing materials.

Practical example

A practical example of chiral materials is in **chiral metamaterials** used in advanced optical devices. These metamaterials are engineered to manipulate electromagnetic waves in novel ways, such as achieving negative refraction or creating invisibility cloaks. The chiral nature of these materials allows them to control light in ways that traditional materials cannot, leading to applications in high-performance lenses, sensors, and communication technologies.

5.11 Applications of SMA

SMAs have a wide range of applications due to their unique ability to return to a pre-defined shape when heated [42]. This SME, along with superelasticity, makes SMAs valuable in various fields.

Applications of SMAs

1. **Medical devices**:
   – **Stents**: SMAs like nitinol are used in stents that can expand and conform to the shape of arteries once deployed, helping to keep blood vessels open.
   – **Guidewires**: In minimally invasive surgeries, SMA guidewires can be flexible yet return to a straight form when needed, improving maneuverability within the body.
2. **Aerospace**:
   – **Actuators**: SMAs are used in aerospace components to adjust surfaces like wing flaps or engine components. Their ability to operate in varying temperatures and return to original shapes makes them ideal for these applications.
3. **Automotive**:
   – **Clutches and brakes**: SMAs can be used in automotive systems for adjusting clutches and brakes, providing reliable and responsive control under different conditions.
4. **Consumer products**:
   – **Eyeglass frames**: Flexible eyeglass frames made from SMAs can be bent and twisted without breaking, returning to their original shape to enhance durability and comfort.
5. **Robotics**:
   – **Grippers**: SMA-based grippers can change their shape in response to temperature changes, allowing robots to handle objects with varying sizes and shapes more effectively.

5.12 Continuum applications of SMA fasteners

SMA fasteners are specialized components that utilize the unique properties of SMAs, particularly their ability to change the shape in response to temperature variations. This functionality allows SMA fasteners to provide innovative solutions in various applications by leveraging their SME and superelasticity.

Continuum applications of SMA fasteners

1. **Aerospace**:
   – **Actuators for adjustable components**: SMA fasteners are used in aerospace systems where components need to adjust or lock into different positions based on temperature changes. For example, they can adjust wing flaps or deploy landing gear by changing their shape in response to temperature variations during flight.
2. **Automotive**:
   – **Thermal-activated fasteners**: In automotive applications, SMA fasteners can be employed in systems requiring thermal activation. They are used in self-locking mechanisms or in components that need to change the configuration when heated, such as engine covers or climate control systems.
3. **Civil engineering**:
   – **Seismic protection systems**: SMA fasteners are utilized in civil engineering for seismic isolation and protection. They can be embedded in structures to provide dynamic response adjustments during an earthquake, enhancing building safety and stability.
4. **Consumer electronics**:
   – **Adjustable housings**: SMA fasteners are used in consumer electronics for adjustable or self-healing housings. For example, in electronic devices with parts that need to adapt to different conditions or repair themselves in response to damage, SMA fasteners offer an elegant solution.
5. **Medical devices**:
   – **Implants and surgical tools**: SMA fasteners are used in medical implants and surgical tools where precise shape changes are required. They can provide adjustable or locking functions in tools or devices that need to adapt to the body's internal environment.

Practical example

A practical example is **SMA-based self-healing aircraft structures**. In aerospace, SMA fasteners are used in the wing or fuselage of an aircraft to automatically adjust the shape or seal damaged sections in response to temperature changes. This ensures structural integrity and improves the longevity of the aircraft components by adapting to operational conditions and repairing minor damage autonomously.

5.13 SMA fibers

SMA fibers are advanced materials that leverage the unique properties of SMAs in a flexible, thread-like form. These fibers exhibit remarkable behaviors due to their inherent SME and superelasticity, making them versatile for various applications. Here is a more detailed look at SMA fibers.

Composition and structure

- **Material composition**: SMA fibers are typically made from alloys such as nitinol (nickel–titanium) or copper-based alloys. These materials are chosen for their ability to undergo phase transformations that result in SME.
- **Manufacturing**: The fibers are manufactured by drawing the SMA material into fine strands or threads. This process involves controlled cooling and heating to achieve the desired properties and dimensions of the fibers.

Properties

- **SME**: SMA fibers can "remember" a specific shape. When deformed at a lower temperature and then heated to a higher temperature, the fibers return to their original shape. This is due to the phase transformation between martensite and austenite states within the alloy.
- **Superelasticity**: At certain temperatures, SMA fibers can undergo significant elastic deformation without permanent damage. They return to their original shape upon removal of the stress, which is useful for applications requiring flexibility and resilience.

Applications

1. **Textiles and wearables**:
   - **Smart clothing**: SMA fibers are woven into fabrics to create garments that adapt to environmental changes. For example, a jacket made with SMA fibers might adjust its fit or insulation properties in response to temperature changes, improving comfort and performance.
2. **Medical devices**:
   - **Implants and surgical tools**: SMA fibers can be used in medical implants, such as stents, where they provide shape adaptability. They can also be incorporated into surgical tools to enhance precision and flexibility during operations.

3. **Structural reinforcement**:
   - **Composite materials**: SMA fibers are integrated into composite materials to provide adaptive reinforcement. For instance, in structural components, these fibers can respond to stress or environmental changes, improving the overall strength and durability of the structure.
4. **Robotics**:
   - **Actuators**: SMA fibers are used in robotic actuators and grippers. Their ability to change shape in response to temperature allows for precise control and adaptability in robotic applications.

Practical example

A practical example of SMA fibers is their use in **adaptive smart textiles** for sportswear. Consider a high-performance jacket made with SMA fibers embedded in its fabric. When the athlete wearing the jacket starts to sweat and their body temperature increases, the SMA fibers in the jacket react by expanding, allowing the jacket to ventilate and adjust its fit for comfort. Conversely, when the temperature drops, the fibers contract, providing better insulation. This dynamic adjustment helps regulate the wearer's temperature and enhances comfort and performance in varying conditions.

SMA fibers' ability to combine flexibility with responsive behavior opens up innovative possibilities across diverse fields, including fashion, medicine, engineering, and robotics.

5.13.1 Reaction vessels

Reaction vessels equipped with SMA fibers are specialized containers used in various chemical, pharmaceutical, and industrial processes. These vessels utilize the unique properties of SMA fibers to enhance performance and adaptability. Here is a brief explanation.

Explanation of SMA fibers in reaction vessels
1. **Temperature-responsive adaptation:**
   - **Temperature regulation**: In reaction vessels, maintaining precise temperature conditions is crucial for optimal chemical reactions. SMA fibers can be incorporated into the vessel's construction to provide temperature-responsive adaptation. When the temperature rises or falls, SMA fibers change shape due to their SME. For instance, if the vessel needs to expand or contract to manage thermal expansion or contraction, SMA fibers will automatically adjust their shape, ensuring consistent internal conditions.

2. **Enhanced sealing and containment:**
   - **Adaptive sealing**: Reaction vessels often deal with high-pressure and high-temperature environments. SMA fibers can be used in seals or gaskets within the vessel. These fibers respond to temperature changes by expanding or contracting, which helps maintain a tight seal. This adaptability is particularly useful in preventing leaks or breaches in the vessel's containment system, ensuring safety and integrity during reactions.
3. **Structural reinforcement and durability:**
   - **Reinforcement**: SMA fibers can be integrated into the vessel's structural materials or lining. These fibers enhance the vessel's strength and resilience by responding to mechanical stresses or thermal changes. For example, during exothermic reactions that generate heat and increase pressure, SMA fibers can provide additional support to the vessel walls, reducing the risk of deformation or failure.
4. **Adaptive functionality:**
   - **Self-repair**: In some advanced designs, SMA fibers could contribute to self-repair mechanisms. If the reaction vessel experiences minor damage or degradation, the fibers could help by adjusting their shape to mitigate the effects of the damage, potentially extending the vessel's service life.
5. **Integration in high-tech processes:**
   - **Precision control**: In highly controlled industrial processes, reaction vessels with SMA fibers offer precise control over internal conditions. For example, in pharmaceutical manufacturing, where maintaining exact temperatures and pressures is critical, SMA fibers can help adjust the vessel's physical properties to ensure optimal conditions for chemical reactions.

Practical example

Chemical processing vessel with SMA fibers:
Consider a chemical processing vessel used in the production of high-value pharmaceuticals. The vessel operates under varying temperature conditions due to exothermic reactions and heat generation.
- **Integration**: SMA fibers are embedded in the vessel's insulation layer and seals.
- **Temperature response**: As the temperature inside the vessel fluctuates, the SMA fibers react by expanding or contracting. This ensures that the vessel maintains a consistent seal, preventing leaks and maintaining pressure.
- **Structural support**: The SMA fibers also provide additional reinforcement to the vessel's walls, helping it withstand the stresses induced by temperature changes and chemical reactions.

By incorporating SMA fibers, the vessel enhances its performance, reliability, and safety. The fiber's ability to adapt to changing conditions ensures that the reaction vessel remains effective and durable throughout its operational life.

5.13.2 Nuclear reactors

In nuclear reactors, SMA fibers are gaining attention for their unique properties, which are highly beneficial for enhancing safety, reliability, and functionality in critical operations. Here is an in-depth explanation of their roles and applications.

Advanced applications of SMA fibers in nuclear reactors

1. **Temperature-responsive components:**
   – **Thermal actuation:** SMA fibers are used in systems where temperature changes can cause a shift in shape, triggering actuation processes. For instance, in control rods or cooling systems, SMA fibers expand or contract based on reactor conditions, allowing for automatic adjustments without external controls. This self-regulating behavior is crucial in maintaining optimal operating conditions in reactors.
2. **Structural health monitoring:**
   – **Crack detection and repair:** SMA fibers can be embedded in critical reactor components, such as pressure vessels or containment structures, for monitoring structural integrity. When small cracks or deformations occur, these fibers can change their shape, either tightening or compressing the material to close gaps and maintain integrity. This self-healing capability is significant in extending the life of reactor components and avoiding costly downtime.
3. **High-temperature seals and gaskets:**
   – **Adaptive sealing solutions:** Reactor environments involve extreme conditions with significant thermal cycling. Conventional seals and gaskets may deteriorate under these conditions. SMA fibers integrated into seals offer better resilience by automatically adjusting to maintain tight seals, even under fluctuating temperatures. This capability is particularly valuable in coolant systems and pressure boundaries, where leak prevention is critical.
4. **Vibration and shock absorption:**
   – **Dynamic damping systems:** Nuclear reactors experience vibrations from both operational sources (e.g., pumps and turbines) and external events (e.g., earthquakes). SMA fibers in damping systems can adapt to these dynamic conditions by altering their stiffness and damping properties. These adaptive systems reduce wear and tear on mechanical components, lower noise, and improve reactor stability.
5. **Fail-safe emergency systems:**
   – **Automated shutdown mechanisms:** In case of abnormal temperature spikes or pressure changes, SMA fibers can be integrated into emergency valves or

containment systems. Upon reaching a certain threshold, these fibers activate emergency protocols by changing the shape, such as sealing off dangerous areas or activating pressure release systems. This rapid response can help prevent catastrophic failures.

Practical example

Adaptive control rod mechanisms

In some advanced reactor designs, SMA fibers are used in control rods that regulate fission. As the temperature in the reactor core rises, SMA fibers embedded in the control rod mechanisms change the shape, adjusting the rod position to either slow down or accelerate the reaction. This adaptive control provides an exact way to manage the reactor's output and safety, responding automatically to changes in operating conditions.

Safety valves with SMA fibers

In a nuclear reactor, safety valves that regulate the coolant flow can be equipped with SMA fibers. During normal operations, these fibers ensure a tight seal. If the temperature exceeds a critical level, the fibers activate, adjusting the valve's position to release pressure or isolate sections of the reactor, preventing potential accidents.

5.13.3 Chemical plants

In chemical plants, SMA fibers are used to enhance safety, efficiency, and functionality in various critical applications. Here is a brief explanation of their roles:

SMA fibers in chemical plants
1. **Leak prevention and sealing solutions:**
   – **Adaptive seals and gaskets:** SMA fibers are integrated into seals and gaskets that operate in high-temperature and corrosive environments common in chemical plants. The SME allows these fibers to respond to temperature changes, ensuring tight seals even under fluctuating conditions. This adaptability is crucial in preventing leaks in pipelines, reactors, and storage tanks where volatile chemicals are handled.
2. **Vibration control and stability:**
   – **Damping systems:** Chemical plants often have heavy machinery and rotating equipment that generate significant vibrations. SMA fibers can be embedded in damping systems to reduce vibrations and mechanical stresses. The fibers adjust their stiffness based on environmental conditions, improving stability and reducing wear on equipment.

3. **Temperature-responsive valves and actuators:**
    – **Smart valves:** SMA fibers are used in temperature-sensitive valves and actuators that automatically adjust flow rates or shut down systems in response to temperature changes. For instance, in safety-critical applications, SMA-actuated valves can automatically close when a certain temperature threshold is exceeded, preventing dangerous reactions or leaks.
4. **Emergency safety systems:**
    – **Fail-safe mechanisms:** In case of abnormal temperature rises or pressure changes, SMA fibers can be designed to activate emergency shutdown systems. For example, they can trigger the closing of safety valves or the release of pressure in chemical reactors, helping to prevent accidents.

Practical example

Automatic temperature-controlled valves
In a chemical plant, SMA fibers are integrated into automatic control valves used in heat exchangers. These valves adjust the flow rates based on real-time temperature changes. If the temperature exceeds safe levels, the SMA fibers contract, closing the valve to prevent overheating or thermal runaway reactions. This application enhances both safety and process efficiency in chemical operations.

5.14 Microrobot actuated by SMA

A microrobot actuated by SMAs uses the unique properties of SMA to achieve movement and functionality in small-scale robotic systems. These microrobots are designed for applications requiring precise and compact actuation, such as in medical devices, micromanipulation, and exploration of small or confined spaces.

Working principle

The SMA in a microrobot functions as an actuator by exploiting its ability to change the shape in response to temperature changes. When the SMA wire or film is heated (usually through an electric current), it undergoes a phase transformation from its martensitic (low-temperature) state to its austenitic (high-temperature) state, causing it to contract. This contraction generates movement, allowing the microrobot to perform specific tasks such as gripping, walking, or crawling. When the SMA cools, it returns to its original shape, completing the cycle.

Components

The components of a microrobot actuated by SMA are designed to optimize its movement and functionality:

1. **SMA actuator:** This is the key element driving the robot's movement. Typically made of SMA wires, springs, or thin films, this component changes the shape when heated, causing the robot to move. The actuator contracts when heated and returns to its original form when cooled, providing controlled and reversible motion.

2. **Control system:** This system manages the actuation process by providing electrical signals to heat the SMA. The control system precisely regulates the amount of current or voltage needed to trigger the phase transformation, ensuring smooth and accurate movements.

3. **Structural frame:** The lightweight and compact body of the microrobot, made from materials such as polymers or metals. The frame provides support, housing the SMA actuator and control system while maintaining a design suited for small-scale operations, such as navigation in confined spaces or delicate tasks.

Applications

– **Medical devices:** Microrobots actuated by SMA are used in minimally invasive surgeries or for targeted drug delivery. They can navigate through blood vessels or other small pathways in the body.

– **Exploration:** In inaccessible or hazardous environments, such as inside machinery or confined spaces, microrobots can be used for inspection and maintenance.

– **Micromanipulation:** This is used in research labs for handling tiny objects or performing delicate operations in microengineering.

Practical example

One example is a microrobot designed for vascular surgery. This robot, actuated by SMA wires, can move through arteries and perform tasks such as removing blockages or delivering therapeutic agents. The robot's compact size and precise actuation make it ideal for such delicate medical applications.

5.15 SMA memorization process

The SMA memorization process for satellite antenna applications involves programming the alloy to "remember" a specific shape, which can be recovered upon activation. This process is crucial in deploying compact, foldable satellite antennas in space.

Memorization process

1. **Shape programming:** The SMA material is first heated above its transformation temperature (austenitic phase) and deformed into the desired "memorized" shape (e.g., the fully extended state of an antenna).
2. **Cooling in deformed state:** The material is then cooled while holding this shape, transitioning into the martensitic phase, where it becomes more malleable.
3. **Temporary folding:** Once cooled, the SMA can be deformed into a compact, folded state for storage or launch. The material "remembers" the original programmed shape despite being folded.
4. **Activation and deployment:** In orbit, an electrical current or another heat source is applied to the SMA, raising its temperature back above the transformation threshold. The SMA reverts to its memorized shape, fully deploying the antenna.

Application in satellite antennas

In satellite systems, SMA antennas are initially stored in a compact form during launch. Once in space, the SMA is activated, causing the antenna to unfold and extend automatically to its operational state. This capability allows for efficient use of space in the launch vehicle and reliable deployment of large structures once in orbit.

5.16 SMA blood clot filter

An SMA blood clot filter is a medical device designed to prevent blood clots from traveling to vital organs like the lungs, where they can cause life-threatening conditions such as pulmonary embolism. The filter, typically made from an SMA-like nitinol (a nickel–titanium alloy), is inserted into the inferior vena cava, a large vein that carries blood from the lower body to the heart.

Working principle

1. **Shape programming:** The SMA filter is preprogrammed with its functional shape (expanded filter state) at high temperatures. This shape is "memorized" by the alloy.
2. **Insertion in a collapsed form:** For implantation, the filter is compressed into a small, compact shape and loaded into a catheter. In this form, it can be easily inserted into the patient's vein through a minimally invasive procedure.

3. **Deployment and self-expansion:** Once in the desired position within the vein, the filter is released from the catheter. As it warms to the body temperature, the SMA material undergoes a phase transformation, expanding back into its original, preprogrammed shape. The filter's legs or struts expand and anchor themselves against the vein walls, creating a mesh that traps blood clots while allowing normal blood flow.

Benefits

– **Minimally invasive:** The use of SMA allows for easy insertion and deployment with minimal discomfort.
– **Self-expanding and adaptive:** The SMA material can conform to different vein sizes, ensuring a secure fit and effective clot filtering.
– **Durable and reliable:** The shape-memory properties provide consistent performance over time, making the filter effective for long-term use.

5.17 Impediments to applications of SMA

While SMAs offer promising applications across various fields, some several impediments and challenges limit their widespread use. These include:

1. **High material costs**
   – SMAs, especially alloys like nitinol (nickel–titanium), are expensive to produce. The precise alloying process and the need for high-quality raw materials contribute to the high costs, limiting their use in cost-sensitive applications.
2. **Fatigue and durability issues**
   – Repeated cycling between phases (austenite and martensite) can lead to material fatigue, reducing the SMA's effectiveness over time. This is particularly problematic in applications requiring constant or high-frequency actuation.
3. **Limited transformation temperatures**
   – The transformation temperature range for SMAs is often narrow, restricting their use to specific environments. For example, if the operational environment's temperature fluctuates beyond the SMA's programmed transformation range, the alloy might not perform as intended.
4. **Complex manufacturing and processing**
   – Manufacturing SMAs with precise control over their properties (e.g., transformation temperatures and stress–strain characteristics) is challenging. Small inconsistencies in processing can lead to variations in performance, making quality control difficult.

5. **Low efficiency and limited actuation force**
   - While SMAs can produce substantial strain, the force generated may be insufficient for certain high-load applications. Additionally, the efficiency of energy conversion is relatively low, with significant energy loss in the form of heat.
6. **Slow response time**
   - The heating and cooling processes required for actuation are relatively slow, which limits the use of SMA in applications where rapid responses are critical.
7. **Corrosion and degradation**
   - Depending on the environment, SMAs can be susceptible to corrosion, especially in biomedical or marine applications. Protective coatings or alloy modifications are often required, which add to the complexity and cost.
8. **Design and control complexity**
   - Implementing SMA-based systems requires precise control mechanisms to ensure accurate actuation and recovery. Developing control systems that manage the thermal activation and response of SMAs adds to the engineering challenge.
9. **Limited lifespan in certain applications**
   - In applications involving continuous stress or high strain, the functional lifespan of SMAs can be shorter due to cumulative fatigue or creep.
10. **Compatibility issues**
    - Integrating SMAs with other materials or components in a system can be complex due to differences in thermal expansion, mechanical properties, or corrosion resistance.

5.18 SMA plastics

Shape-memory plastics, often referred to as shape-memory polymers (SMPs), are a class of smart materials that exhibit the ability to "remember" and return to their original shape after deformation when exposed to a specific stimulus, such as temperature, light, or magnetic field. Unlike traditional SMAs, SMPs are lightweight, flexible, and can undergo much larger deformations.

Key features

- **Lightweight and flexible**: SMPs are much lighter than metallic SMAs, making them ideal for applications where weight is a critical factor.
- **Large deformations**: SMPs can undergo significant shape changes (up to 200–400% strain) compared to SMAs, which are limited to around 5–8%.

- **Customizable trigger**: The shape recovery in SMPs can be triggered by various stimuli, including heat, light, electricity, or even magnetic fields.
- **Low cost**: SMPs are typically less expensive to produce than SMAs, making them more accessible for commercial applications.

Applications

1. **Biomedical devices**: SMPs are used in stents, sutures, and other medical implants that can change the shape within the body for minimally invasive surgeries.
2. **Deployable structures in aerospace**: SMPs are employed in self-deploying structures such as antennas or solar panels that unfold in space.
3. **Textiles and wearables**: SMPs are incorporated into adaptive clothing that changes the shape or fit in response to body temperature or environmental conditions.
4. **Smart packaging**: Packaging materials that can change the shape or size depending on the storage conditions, improving efficiency and shelf life.

Practical example

One practical application is in minimally invasive surgery. SMPs can be used in medical stents that are initially small and flexible for easy insertion into the body. Once in position, they can be triggered (e.g., by body heat) to expand and provide structural support to blood vessels or other tissues, making them more adaptable and less intrusive than traditional metal stents.

In the context of SMPs, **primary molding** and **secondary molding** are crucial processes that determine how the material achieves and recovers its shape during applications.

Primary molding:

Primary molding is the initial process where the SMP is shaped into its "memorized" or permanent form. This process typically involves:

- **Molding technique:** The polymer is heated above its transition temperature and then molded into the desired permanent shape.
- **Setting the shape:** After molding, the polymer is cooled, fixing the shape in memory. This "memorized" shape is the form the material will return to when activated by a trigger (e.g., heat).

For example, if the SMP is designed for a stent, during primary molding, it is formed into its expanded configuration (permanent shape), which it will revert to when used.

Secondary molding:

Secondary molding involves deforming the SMP into a temporary shape, which can be reverted back to the original "memorized" shape when triggered. Steps include:

- **Heating above the transition temperature:** The material is reheated to make it malleable.
- **Deformation:** The SMP is deformed into a temporary shape, often by stretching, compressing, or bending.
- **Cooling to fix the temporary shape:** The polymer is cooled below its transition temperature while held in the temporary shape. It retains this shape until it is reheated.

In applications like deployable aerospace structures, secondary molding allows the structure to be compacted for transport. When triggered (e.g., by sunlight heating the material), it returns to its larger, functional shape.

Practical example

In packaging technology, primary molding might involve creating an SMP in a compact shape for easy transportation. Secondary molding is then applied to fold or compress the package. Once delivered, exposure to heat or light could expand the package back to its full size, thanks to the SME.

5.19 Potential applications of SMA plastics

SMA plastics, also known as shape-memory polymers, have versatile applications across various industries due to their ability to change shape when exposed to specific stimuli like heat, light, or magnetic fields. Some potential applications include:

1. **Medical devices:**
   - **Stents and implants:** SMPs can be used in minimally invasive surgeries. Stents made from SMPs can be compressed for easy insertion and then expand to their functional shape inside the body.
   - **Orthopedic braces and splints:** SMPs can be custom-fitted to a patient's anatomy and adjusted when needed through heat activation.
   - **Drug delivery systems:** SMPs can encapsulate drugs and release them on-demand when exposed to a trigger, allowing controlled drug release.
2. **Textiles and wearables:**
   - **Self-adjusting garments:** Clothing that changes the shape based on body temperature or external conditions, providing enhanced comfort and functionality.
   - **Smart bandages:** SMPs can be used in bandages that tighten or loosen automatically based on temperature changes for optimized wound healing.

3. **Aerospace and automotive:**
   - **Deployable structures:** SMPs can be used in space applications for deployable structures, such as antennas, solar panels, or wings, which are compact during launch and expand to their full size in orbit.
   - **Adaptive components:** In vehicles, SMPs can be used in components that adjust aerodynamics, such as spoilers or vents, to optimize performance based on driving conditions.
4. **Consumer electronics:**
   - **Flexible electronics:** SMPs can be incorporated into flexible screens or cases that change the shape for better user experience or protection.
   - **Responsive buttons and switches:** Devices with buttons that emerge or retract based on usage patterns.
5. **Robotics:**
   - **Soft robotics:** SMPs are ideal for creating soft, adaptive robotic components that change the shape and stiffness to perform complex tasks.
   - **Actuators and grippers:** SMPs can be used in grippers that adjust their shape to hold different objects, improving flexibility and precision.
6. **Packaging and logistics:**
   - **Reconfigurable packaging:** Packaging materials that change the shape to fit different products or expand when needed, reducing storage space and enhancing protection.
   - **Temperature-responsive seals:** SMPs can be used for packaging seals that adjust based on temperature to ensure freshness or tamper-proofing.

5.20 Summary

SMAs are unique materials that can revert to a predetermined shape when subjected to specific stimuli such as temperature changes. These materials, such as nickel alloys and titanium-based nitinol, have characteristics like SME and superelasticity due to martensitic and austenitic phase transformations. The thermoelastic martensitic transformation is key to their behavior. Cu-based SMAs are also explored for specialized applications, including aerospace. SMAs find their use in various fields such as chirality in advanced materials, fasteners, SMA fibers in reaction vessels, and nuclear reactors. Microrobots and satellite antennas are actuated using SMAs, while blood clot filters showcase the medical potential. Challenges remain, including design impediments and manufacturing limitations such as primary and secondary molding processes. SMA plastics, a growing field, offer innovative applications, such as adaptive medical devices, reconfigurable packaging, and smart textiles, broadening the scope of these versatile materials.

Review questions

1. What is the basic principle behind SMAs?
2. How do SMAs revert to their original shape?
3. What are the main characteristics of nickel alloys used in SMAs?
4. Describe the composition and applications of nitinol.
5. How do martensitic transformations enable the SME in SMAs?
6. What is the role of austenitic transformations in SMAs?
7. Explain the thermoelastic martensitic transformation in SMAs.
8. What are the advantages of Cu-based SMAs in aerospace applications?
9. How do chiral materials relate to SMAs?
10. List the key applications of SMAs.
11. What are the benefits of using SMAs in fasteners for structural applications?
12. How are SMA fibers utilized in high-temperature environments?
13. Describe the role of SMA fibers in reaction vessels.
14. What are the functions of SMA fibers in nuclear reactors?
15. How do SMA fibers improve the efficiency of chemical plants?
16. Explain the working principle of a microrobot actuated by SMA.
17. What are the components involved in an SMA-based microrobot?
18. How is the SMA memorization process used in satellite antenna applications?
19. What is the function of an SMA blood clot filter?
20. What are some key impediments to the widespread application of SMAs?
21. How does primary molding differ from secondary molding in SMA plastics?
22. What are potential applications of SMA plastics in medical devices?
23. Discuss how SMA plastics can be used in adaptive packaging solutions.
24. What are the main characteristics of chiral materials used in SMA-based applications?
25. How do Cu–Al–Ni alloys contribute to high-temperature SMA applications?
26. What are the primary challenges faced in SMA-based microdevices?
27. Explain the role of martensitic transformations in reversible shape recovery.
28. How does temperature influence the behavior of thermoelastic martensitic transformations?
29. What makes nitinol a widely used material in biomedical applications?
30. How do SMA-based actuators function in real-world scenarios?
31. Describe the role of SMAs in continuum applications, specifically fasteners.
32. What factors impact the performance of SMA fibers in harsh environments?
33. Explain the significance of SMA fibers in high-stress applications like chemical plants.
34. How do SMA fibers enhance safety in nuclear reactor systems?
35. Discuss the importance of SMA memorization processes in aerospace technology.
36. How do SMAs contribute to medical innovations like blood clot filters?
37. What are the processing techniques for molding SMA plastics?

38. Compare the primary and secondary molding methods for SMA components.
39. What are the potential advantages of using SMA plastics in smart textiles?
40. Explain the challenges in designing microrobots actuated by SMA.
41. What makes chiral materials important in developing next-generation SMAs?
42. How do thermal cycles impact the stability of martensitic and austenitic phases?
43. What are the criteria for selecting SMAs for specific engineering applications?
44. How does the Cu-based SMA's transformation temperature influence its application?
45. Explain the advantages of using SMA-based fasteners in aerospace structures.
46. How do SMA fibers aid in structural health monitoring systems?
47. What role do SMA actuators play in miniaturized robotic devices?
48. Discuss the thermal management challenges in SMA-based applications.
49. What are the environmental considerations in using SMA plastics?
50. How do SMA fibers provide enhanced control in precision actuation systems?

Authors' bibliography

Kaushik Kumar holds his **Ph.D.** in **engineering** from **Jadavpur University, India**, **MBA** in **marketing management** from **Indira Gandhi National Open University, India**, and **Bachelor of Technology** from **Regional Engineering College (Now National Institute of Technology), Warangal, India**. For 11 years, he worked in a manufacturing unit of global repute. He is currently working as a professor in the Department of Mechanical Engineering, Birla Institute of Technology, Mesra, Ranchi, India. He has 23 years of teaching and research experience. His research interests include composites, optimization, nonconventional machining, CAD/CAM, rapid prototyping, and quality management systems toward **product development for societal and industrial usage** and has received **32 patents** for them. He has published **55+ books** (including **31 edited book** volumes) (referred to as **text books and reference books** by **40+ universities/institutes** in their academic curriculum), **90+ book chapters**, and **200+ research papers** in peer-reviewed reputed national and international journals. Kaushik has also served as **editor-in-chief, series editor, guest editor, editor, editorial board member**, and **reviewers** for international and national journals. He has been felicitated with many awards and honors, including **Distinguished Alumnus Award for Professional Excellence 2023 under Academic and Research** from his alma mater **National Institute of Technology, Warangal, India**. He has also received sponsored research and consultancy projects of **more than 1 crore** from Government of India and abroad. Kaushik has delivered expert lectures as keynote speaker at international and national conferences, resource person at various workshops, FDPs and short-term courses. He has guided many students of doctoral, masters, and undergraduate programs of his home and other institutions in India and abroad. He has also served as reviewer and examiner of doctoral and masters dissertation for institutes in India and abroad.

Chikesh Ranjan is a highly accomplished Mechanical Engineer currently serving as a Project Engineer for Project SwaYaan – Capacity Building in Drone/UAS at the Department of Mechanical Engineering, NIT Rourkela. He holds a PhD in Mechanical Engineering from the Department of Mechanical Engineering, Birla Institute of Technology, Mesra, Ranchi, Jharkhand, India. Chikesh also possesses a Master of Engineering (M.E) degree in Mechanical Engineering, specializing in the Design of Mechanical Equipment, from the same institution.

His academic journey began with a Bachelor of Engineering (B.E) in Mechanical Engineering from Marathwara Institute of Technology, Aurangabad, Maharashtra, India, where he graduated with First Class with distinction. Chikesh's innovative contributions have been recognized through the publication of 15 journal papers, 10 authored books, and 13 book chapters. He has actively organized 10 workshops, coordinated 13 events, and been involved in 18 events overall. Additionally, he has participated in 31 training/FDP programs and attended 34 national/international workshops, seminars, and webinars.

Chikesh has been granted 10 design patents and 7 foreign patents, demonstrating his dedication to advancing mechanical design and technology. With 9 years of teaching and research experience of global repute, his areas of teaching and research interest include composites, non-conventional machining, CAD/CAM, and robotics. He is a proud member of the Institution of Engineers, India, and has received 7 prestigious awards and honors for his remarkable contributions to the field of Mechanical Engineering.

https://doi.org/10.1515/9783111379623-006

Bibliography

[1] Suhag, D. (2024). Biomaterials in Oncology. In Handbook of Biomaterials for Medical Applications, Volume 2: Applications (pp. 171–204). Springer Nature Singapore, Singapore. https://doi.org/10.1007/978-981-97-5906-4_6.

[2] Ebrahimi, F. & Ahari, M. F. (2024). Mechanics of Active Materials 234–268. In Silberschmidt, V. (Ed.) Comprehensive Mechanics of Materials. 1st Edition (ISBN 9780323906470), Elsevier Inc., USA. https://doi.org/10.1016/B978-0-323-90646-3.00043-5.

[3] Schwartz, M. (2002). Encyclopaedia of Materials, Parts and Finishes. 2nd Edition (ISBN 9780429133183), CRC Press, Boca Raton, FL. https://doi.org/10.1201/9781420017168.

[4] Gupta, M. N. (2024). Smart Systems in Biotechnology. 1st Edition, CRC Press, Boca Raton, FL. https://doi.org/10.1201/9781003328919.

[5] Alhazmi, I. (2024). Punching Shear Strengthening of Heat-Damaged Normal and High-Strength Reinforced Concrete Flat Slabs Using NSM-FRP Strips and Ropes (Doctoral dissertation, The University of Akron).

[6] Kumar, A., Dogra, N., Bhatia, S. & Sidhu, M. S. (Eds.). (2024). Handbook of Intelligent and Sustainable Smart Dentistry: Nature and Bio-inspired Approaches, Processes, Materials, and Manufacturing. 1st Edition, CRC Press, Boca Raton, FL. https://doi.org/10.1201/9781003404934.

[7] Chen, Q., Kalpoe, T. & Jovanova, J. (2024). Design of mechanically intelligent structures: review of modeling stimuli-responsive materials for adaptive structures. Heliyon, 10(14). https://doi.org/10.1016/j.heliyon.2024.e34026.

[8] Ponnamma, D., Sadasivuni, K. K., Cabibihan, J. J. & Al-Maadeed, M. A. A. (2017). Smart Polymer Nanocomposites, Springer Series on Polymer and Composite Materials. Springer, Switzerland. https://doi.org/10.1007/978-3-319-50424-7.

[9] Gunjal, P. R., Jondhale, S. R., Mauri, J. L. & Agrawal, K. (2024). Internet of Things: Theory to Practice. 1st Edition, CRC Press, Boca Raton, FL. https://doi.org/10.1201/9781003282945.

[10] Flatau, A. B. & Chong, K. P. (2002). Dynamic Smart Material and Structural Systems. Engineering Structures, 24(3), 261–270. https://doi.org/10.1016/S0141-0296(01)00093-1.

[11] Zarzar, L. D. & Aizenberg, J. (2014). Stimuli-responsive chemomechanical actuation: A hybrid materials approach. Accounts of Chemical Research, 47(2), 530–539. https://doi.org/10.1021/ar4001923.

[12] Rahaman, M. N. & Mao, J. J. (2005). Stem cell-based composite tissue constructs for regenerative medicine. Biotechnology and Bioengineering, 91(3), 261–284. https://doi.org/10.1002/bit.20292.

[13] Behera, A., Nayak, A. K., Mohapatra, R. K. & Rabaan, A. A. (Eds.). (2024). Smart Micro-and Nanomaterials for Pharmaceutical Applications. 1st Edition, CRC Press, Boca Raton, FL. https://doi.org/10.1201/9781003468431.

[14] Shukla, S. K. (2024). Thermal Evaluation of Indoor Climate and Energy Storage in Buildings. 1st Edition, CRC Press, Boca Raton, FL. https://doi.org/10.1201/9781003415695.

[15] Hayashi, K., Shindo, Y. & Narita, F. (2003). Displacement and polarization switching properties of piezoelectric laminated actuators under bending. Journal of Applied Physics, 94(7), 4603–4607. https://doi.org/10.1063/1.1603963.

[16] Lantada, A. D. & Morgado, P. L. (2011). Active Materials in Medical Devices. In Lantada, A. D. (Ed.) Handbook of Active Materials for Medical Devices: Advances and Applications, 57–90 Jenny Stanford Publishing, NY. https://doi.org/10.1201/b11170.

[17] Kishi, T. (2003). The forty-eighth Honda Memorial Lecture nondestructive evaluation and smart materials. Materials Transactions, 44(8), 1546–1552. https://doi.org/10.2320/matertrans.44.1546.

[18] Sharif, M. (2023). Advances in Micropump Technology: Harnessing Electrorheological and Magnetorheological Fluids for Optimal Performance. http://dx.doi.org/10.31219/osf.io/xagjk

https://doi.org/10.1515/9783111379623-007

[19]   Gołdasz, J. & Sapiński, B. (2015). Insight into Magnetorheological Shock Absorbers. Springer Cham, Springer International Publishing Switzerland. https://doi.org/10.1007/978-3-319-13233-4.

[20]   Cohen, E. D. & Gutoff, E. B. (2000). Coating Processes, Survey. In Kirk-Othmer Encyclopedia of Chemical Technology. Wiley Online Library. https://doi.org/10.1002/0471238961. 1921182203150805.a01.

[21]   Smith, A. (2023). Advancements in Pump Technology: Solid State Pump Utilizing Electro-Rheological Fluid. https://doi.org/10.31219/osf.io/73hav

[22]   Behera, A. (2021). Advanced Materials: An Introduction to Modern Materials Science. Springer Cham, Springer Nature Switzerland AG 2022. https://doi.org/10.1007/978-3-030-80359-9.

[23]   Rafiee, M., Nitzsche, F. & Labrosse, M. (2017). Dynamics, vibration and control of rotating composite beams and blades: a critical review. Thin-Walled Structures, 119, 795–819. https://doi.org/10.1016/j.tws.2017.06.018.

[24]   Nikolakopoulos, P. G. & Papadopoulos, C. A. (1998). Controllable high speed journal bearings, lubricated with electro-rheological fluids. An analytical and experimental approach. Tribology International, 31(5), 225–234. https://doi.org/10.1016/S0301-679X(98)00025-5.

[25]   Broze, G. (Ed.). (1999). Handbook of Detergents, Part A: Properties. 1st Edition, CRC Press, Boca Raton, FL. https://doi.org/10.1201/9780367803254.

[26]   Gunjal, P. R., Jondhale, S. R., Mauri, J. L. & Agrawal, K. (2024). Internet of Things: Theory to Practice. 1st Edition, CRC Press, Boca Raton, FL. https://doi.org/10.1201/9781003282945.

[27]   Enuka, A. M. (2024). Manufacturing of Aligned Electrospun Nanofiber Yarns for Their Enhanced Mechanical and Piezoelectric Properties (Master's thesis, Department of Chemical Engineering, Rowan University). https://rdw.rowan.edu/etd/3259

[28]   Ali, M. R. R., Tigno, S. D. & Caldona, E. B. (2024). Piezoelectric approaches to organic polymeric materials. Polymer International, 73(3), 176–190. https://doi.org/10.1002/pi.6601.

[29]   Kannan, S. (2023). Acoustic-emission Monitoring of Lap Joint Fatigue Cracks (Master's thesis, Department of Mechanical Engineering, University of South Carolina). https://scholarcommons.sc.edu/etd/7635

[30]   Vinoy, K. J., Ananthasuresh, G. K., Pratap, R. & Krupanidhi, S. B. (Eds.). (2014). Micro and Smart Devices and Systems (p. 184) Springer India. https://doi.org/10.1007/978-81-322-1913-2.

[31]   Taylor, J. & Huang, Q. (2020). CRC Handbook of Electrical Filters. 1st Edition, CRC Press, Boca Raton, FL. https://doi.org/10.1201/9781003069201.

[32]   Fulay, P. & Lee, J. K. (2016). Electronic, Magnetic, and Optical Materials. 2nd Edition, CRC Press, Boca Raton, FL. https://doi.org/10.1201/9781315371870.

[33]   Addington, M. & Schodek, D. (2012). Smart Materials and Technologies in Architecture. Routledge, London. https://doi.org/10.4324/9780080480954.

[34]   Shankar, K. (2024). Design and Manufacturing of a Shape Memory Alloy-Based Actuator. (Master's thesis, Department of Mechanical Engineering, Arizona State University). https://keep.lib.asu.edu/items/193483

[35]   Srivastava, M., Rathee, S., Maheshwari, S. & Kundra, T. K. (2019). Additive Manufacturing: Fundamentals and Advancements. 1st Edition, CRC Press, Boca Raton, FL. https://doi.org/10.1201/9781351049382.

[36]   Shah, S. I. H. & Lim, S. (2024). RF advancements enabled by smart shape memory materials in the microwave regime: a state-of-the-art review. Materials Today Physics, 44, 101435. https://doi.org/10.1016/j.mtphys.2024.101435.

[37]   Hartl, D. J., Mooney, J. T., Lagoudas, D. C., Calkins, F. T. & Mabe, J. H. (2009). Use of a Ni60Ti shape memory alloy for active jet engine chevron application: II. Experimentally validated numerical analysis. Smart Materials and Structures, 19(1), 015021. https://doi.org/10.1088/0964-1726/19/1/015021.

[38] Shahi, K. & Ramachandran, V. (2024). Shape memory effect in polymers: Experiment and theory. Comprehensive Materials Processing. 2nd Edition, 12, 485–508. https://doi.org/10.1016/B978-0-323-96020-5.00261-2.

[39] Fisher, P. E. (2005). Selection of Engineering Materials and Adhesives. 1st Edition, CRC Press, Boca Raton, FL. https://doi.org/10.1201/b15905.

[40] Leng, J. & Du, S. (Eds.). (2010). Shape-Memory Polymers and Multifunctional Composites. 1st Edition, CRC Press, Boca Raton, FL. https://doi.org/10.1201/9781420090208.

[41] Lu, H., Yu, K., Sun, S., Liu, Y. & Leng, J. (2010). Mechanical and shape-memory behavior of shape-memory polymer composites with hybrid fillers. Polymer International, 59(6), 766–771. https://doi.org/10.1002/pi.2785.

[42] Tomlinson, G. R. & Bullough, W. A. (1998). Smart Materials and Structures: Proceedings of the 4th European and 2nd MIMR Conference. Harrogate, UK, 6–8 July 1998. CRC Press, Boca Raton, FL. https://doi.org/10.1201/9781482268560.

[43] Georges, T., Brailovski, V. & Terriault, P. (2012). Characterization and design of antagonistic shape memory alloy actuators. Smart Materials and Structures, 21(3), 035010. https://doi.org/10.1088/0964-1726/21/3/035010.

Index

https://doi.org/10.1515/9783111379623-008

www.ingramcontent.com/pod-product-compliance
Lightning Source LLC
Chambersburg PA
CBHW081534220326
41598CB00036B/6435